ECO-COMPASS

ECO-COMPASS: Ecological and Multifunctional Composites for Application in Aircraft Interior and Secondary Structures

Special Issue Editors

Xiaosu Yi
Konstantinos Tserpes

MDPI • Basel • Beijing • Wuhan • Barcelona • Belgrade

Special Issue Editors

Xiaosu Yi
University of Nottingham Ningbo China (UNNC)
China

Konstantinos Tserpes
University of Patras
Greece

Editorial Office
MDPI
St. Alban-Anlage 66
4052 Basel, Switzerland

This is a reprint of articles from the Special Issue published online in the open access journal *Aerospace* (ISSN 2226-4310) from 2018 to 2019 (available at: https://www.mdpi.com/journal/aerospace/special_issues/ECO-COMPASS).

For citation purposes, cite each article independently as indicated on the article page online and as indicated below:

LastName, A.A.; LastName, B.B.; LastName, C.C. Article Title. *Journal Name* **Year**, *Article Number*, Page Range.

ISBN 978-3-03897-690-5 (Pbk)
ISBN 978-3-03897-691-2 (PDF)

Contents

About the Special Issue Editors

Xiaosu Yi Earned his Dipl.-Ing. (M.S.) in 1982 and his Dr.-Ing. (Ph.D.) degree in 1986 in Material Engineering at the University of Paderborn, Germany. A full professor in Polymer Materials and Technology since 1987 at Zhejiang University, Hangzhou, China, he joined the Beijing Institute of Aeronautical Materials (BIAM) in 1998, before moving on to the AVIC Composite Center (ACC) in 2010. Since 2017, he has been a Chair professor at the University of Nottingham, Ningbo, China. He is the director of the National Engineering Laboratory of Carbon-Fiber Structural Composites and the Beijing Engineering Laboratory of Green Composites. Fellow of SAMPE (F. SAMPE), Fellow of Aviation Industry of China (AVIC), Academician of APAM (Asia-Pacific Academy of Materials), Chairman of SAMPE China, and an executive council member of the International Conference of Composite Materials (ICCM) and Asia-Pacific Conference of Composite Materials (ACCM). He has, to date, received the SAMPE Fellow Award, the ZhouGuangZhao Award, a KC Wong Fellowship, the Aerospace Laureate Award, the Feng Ru Aerospace Science and Technology Award, and the China Patent Excellence Award. He has published approximately 400 papers, 8 monographs, and approximately 60 invention patents.

Konstantinos Tserpes is a Mechanical and Aeronautical Engineer with a Ph.D. in strength prediction of composite materials. Currently, he is Assistant Professor at the Department of Mechanical Engineering and Aeronautics, University of Patras, Greece. His research interests are in the areas of strength of composite materials, strength of bonded and bolted joints, mechanical behavior of carbon nanotubes, graphene, nanocomposites and nanocrystalline materials, multiscale analysis of materials and structural parts, development of methodologies for relating data from nondestructive testing with numerical strength prediction models, as well as strength prediction of corroded aluminum parts. He has co-edited four international books and published 6 chapters in books, as well as more than 65 papers in journals and more than 75 papers in conference proceedings. He has 20 years of research experience in the areas of aeronautical materials and structures gained through participation in more than 20 national and international research projects.

Editorial

Special Issue "ECO-COMPASS: Ecological and Multifunctional Composites for Application in Aircraft Interior and Secondary Structures"

Xiaosu Yi [1,2,*] and Konstantinos Tserpes [3,*]

[1]   Beijing Institute of Aeronautical Materials, Beijing 100095, China
[2]   Aviation Composite (Beijing) Science and Technology Co., Ltd., Beijing 101300, China
[3]   Laboratory of Technology & Strength of Materials, Department of Mechanical Engineering & Aeronautics, University of Patras, Patras 26500, Greece
*   Correspondence: xiaosu.yi@nottingham.edu.cn (X.Y.); kitserpes@upatras.gr (K.T.)

Received: 12 February 2019; Accepted: 13 February 2019; Published: 13 February 2019

Today, composite aircraft structural parts are mainly made of man-made materials, such as carbon and glass fibres and epoxy resin. Renewable materials, such as natural fibres or bio-sourced resin systems, have not yet found their way into aircraft production. The project ECO-COMPASS [1] aims to evaluate the potential applications of ecologically improved composite materials in the aeronautics sector through an international collaboration between Chinese and European partners. Natural fibres, such as flax and ramie, are used for different types of reinforcements and sandwich cores. The substitution of bisphenol-A based epoxy resins in secondary structures by bio-based epoxy resins is currently under investigation. Adapted material protection technologies aiming to reduce environmental influence and to improve fire resistance are needed to fulfil the demanding safety requirements in aviation. The modelling and simulation of the chosen eco-composites aims to optimize the use of materials while a Life Cycle Assessment aims to prove the ecological advantages compared to the synthetic state-of-the-art materials.

This Special Issue provides selected papers from the project consortium partners. The Special Issue is partially based on the special session entitled "ECO-COMPASS: Ecological and Multifunctional Composites for Application in Aircraft Interior and Secondary Structures" that was organized at the ICCS20 Conference (Paris, France, 4–7 September 2017, https://events.unibo.it/iccs20).

This Special Issue of Aerospace contains nine interesting articles, which cover a wide range of topics from production, experimental characterization and numerical simulation. The paper by Barbara Tse et al. [2] characterizes the flexural behavior and the morphological properties of wet-laid hybrid nonwoven recycled carbon and flax fibre composites in a polylactic acid matrix. Experimental data showed that the flexural properties increased with higher recycled carbon fibres (rCF) content. The intimate mixing of the fibres contributed to a lesser reduction of flexural properties when increasing the flax fibre content. Jens Bachmann et al. [3] measured flexural properties of hybrid epoxy composites reinforced with nonwoven flax and recycled carbon fibres. Experimental results show a potential increase in flexural properties after combining rCF and flax fibre in a nonwoven hybrid. Tserpes and Kora [4] proposed a multiscale modeling approach to simulate crack sensing in polymer fibrous composites by exploiting the interruption of electrically conductive carbon nanotube (CNT) networks. In the special issue, the second paper of a two paper series [4,5] was published. The numerical results highlight the prospect of conductive CNT networks to be used as a localized structural health monitoring technique in carbon fibre reinforced polymers (CFRP) and bio-composites. Wang et al. [6] studied the effect of ramie fabric chemical treatments on the physical properties of thermoset polylactic acid (PLA) composites. It was found that chemical treatments lead to an increase in tensile and flexural strength of PLA composites while they lead to a decrease in water absorption. The authors concluded that the ramie fabric-reinforced PLA composites can meet the standard requirements of aircraft interior structures and have favorable foreground application.

Dong et al. [7] performed cradle-to-gate life cycle assessment (LCA) study to demonstrate the possible advantages of ramie fibre on environmental impacts and to provide fundamental data for the further assessment of ramie fibre reinforced polymers (RFRP) and its structures. Guo et al. [8] studied the effect of plant-fibre paper or silver nanowires-loaded paper interleaves on the electrical conductivity and interlaminar fracture toughness of composites. Experimental data show an increase in electrical conductivity and a decrease in interlaminar fracture toughness. Zhang et al. [9] evaluated the sound absorption performance of flax fibre and its reinforced composite, as well as balsa wood, using the two-microphone transfer function technique with an impedance tube system. The sandwich structure with integrated natural materials was found to provide a superior sound absorption performance compared to the synthetic-materials-based sandwich structure composite. Yi et al. [10] reported on current R&D efforts to develop bio-sourced materials by an international joint project. Novel bio-sourced epoxies and biocomposites were developed, characterized, modified and evaluated in terms of the mechanical property levels. Quasi-structural composite parts were finally trial-manufactured and demonstrated. Finally, the paper of Ramon et al. [11] reviewed recent advances on new bio-based epoxy resins, which were derived from natural oils, natural polyphenols, saccharides, natural rubber and rosin.

The editors of this Special Issue would like to thank the authors for their high-quality contributions and for making this Special Issue a success. Additionally, the editors would like to thank the Aerospace editorial office, in particular Ms. Linghua Ding.

Funding: This project has received funding from the European Union's Horizon 2020 research and innovation programme under grant agreement No 690638.

Conflicts of Interest: The authors declare no conflict of interest.

References

1. Bachmann, J.; Yi, X.; Gong, H.; Martinez, X.; Tserpes, K.; Ramon, E.; Paris, C.; Moreira, P.; Fang, Z.; Li, Y.; et al. Outlook on ecologically improved composites for aviation interior and secondary structures. *CEAS Aeronaut. J.* **2018**, *9*, 533–543. [CrossRef]
2. Tse, B.; Yu, X.; Gong, H.; Soutis, C. Flexural Properties of Wet-Laid Hybrid Nonwoven Recycled Carbon and Flax Fibre Composites in Poly-Lactic Acid Matrix. *Aerospace* **2018**, *5*, 120. [CrossRef]
3. Bachmann, J.; Wiedemann, M.; Wierach, P. Flexural Mechanical Properties of Hybrid Epoxy Composites Reinforced with Nonwoven Made of Flax Fibres and Recycled Carbon Fibres. *Aerospace* **2018**, *5*, 107. [CrossRef]
4. Tserpes, K.; Kora, C. A Multi-Scale Modeling Approach for Simulating Crack Sensing in Polymer Fibrous Composites Using Electrically Conductive Carbon Nanotube Networks. Part II: Meso- and Macro-Scale Analyses. *Aerospace* **2018**, *5*, 106. [CrossRef]
5. Tserpes, K.; Kora, C. A multi-scale modeling approach for simulating crack sensing in polymer fibrous composites using electrically conductive carbon nanotube networks. Part I: Micro-scale analysis. *Comput. Mater. Sci.* **2018**, *154*, 530. [CrossRef]
6. Wang, C.; Ren, Z.; Li, S.; Yi, X. Effect of Ramie Fabric Chemical Treatments on the Physical Properties of Thermoset Polylactic Acid (PLA) Composites. *Aerospace* **2018**, *5*, 93. [CrossRef]
7. Dong, S.; Xian, G.; Yi, X.-S. Life Cycle Assessment of Ramie Fiber Used for FRPs. *Aerospace* **2018**, *5*, 81. [CrossRef]
8. Guo, M.; Yi, X. Effect of Paper or Silver Nanowires-Loaded Paper Interleaves on the Electrical Conductivity and Interlaminar Fracture Toughness of Composites. *Aerospace* **2018**, *5*, 77. [CrossRef]

9.   Zhang, J.; Shen, Y.; Jiang, B.; Li, Y. Sound Absorption Characterization of Natural Materials and Sandwich Structure Composites. *Aerospace* **2018**, *5*, 75. [CrossRef]

10.   Yi, X.-S.; Zhang, X.; Ding, F.; Tong, J. Development of Bio-Sourced Epoxies for Bio-Composites. *Aerospace* **2018**, *5*, 65. [CrossRef]

11.   Ramon, E.; Sguazzo, C.; Moreira, P.M.G.P. A Review of Recent Research on Bio-Based Epoxy Systems for Engineering Applications and Potentialities in the Aviation Sector. *Aerospace* **2018**, *5*, 110. [CrossRef]

Article

Flexural Properties of Wet-Laid Hybrid Nonwoven Recycled Carbon and Flax Fibre Composites in Poly-Lactic Acid Matrix

Barbara Tse [1,*], Xueli Yu [1], Hugh Gong [1] and Constantinos Soutis [1,2]

[1]   School of Materials, The University of Manchester, Manchester M13 9PL, UK;
xueli.yu@postgrad.manchester.ac.uk (X.Y.); hugh.gong@manchester.ac.uk (H.G.);
constantinos.soutis@manchester.ac.uk (C.S.)

[2]   Aerospace Research Institute, The University of Manchester, Manchester M13 9PL, UK

*   Correspondence: barbara.tse@manchester.ac.uk; Tel.: +44-161-306-2868

Received: 3 September 2018; Accepted: 5 November 2018; Published: 15 November 2018

Abstract: Recycling carbon fibre is crucial in the reduction of waste from the increasing use of carbon fibre reinforced composites in industry. The reclaimed fibres, however, are usually short and discontinuous as opposed to the continuous virgin carbon fibre. In this work, short recycled carbon fibres (rCF) were mixed with flax and poly-lactic acid (PLA) fibres acting as the matrix to form nonwoven mats through wet-laying. The mats were compression moulded to produce composites with different ratios of rCF and flax fibre in the PLA matrix. Their flexural behaviour was examined through three-point-bending tests, and their morphological properties were characterised with scanning electron and optical microscopes. Experimental data showed that the flexural properties increased with higher rCF content, with the maximum being a flexural modulus of approximately 14 GPa and flexural strength of 203 MPa with a fibre volume fraction of 75% rCF and 25% flax fibre. The intimate mixing of the fibres contributed to a lesser reduction of flexural properties when increasing the flax fibre content.

Keywords: hybrid composite; eco-composite; nonwoven; recycled carbon fibre; flax fibre; poly-lactic acid; wet-laying

---

1. Introduction

Carbon fibres (CF) are the preferred reinforcement material for polymer composites because they combine high strength and low weight compared to metallic or ceramic fibres [1]. The global demand for CF rose from 33,000 t in 2010 to 72,000 t in 2017 and is expected to grow at a rate of 9–12% for the next five years [2]. This increase of CF products comes with the increasing levels of waste resulting from expired prepregs, production cut-offs, testing materials, and end-of-life components from the aeronautics, automotive, and wind industries [3].

Due to the energy-intensive manufacturing of virgin CF [4], it also becomes more economical to recover and reuse CF. In the ideal case, recycled carbon fibres (rCF) would replace virgin fibres to save more energy and reduce production cost. However, fibre damage and resin residues from the recycling process decreases the fibre tensile strength to around 80% of the virgin fibres, depending on the recycling method, through thermal, mechanical, and chemical recycling treatments [5–7]. For mechanical recycling, the composites are crushed and sieved, which results in a powdered form that can be further used for injection moulding. In chemical recycling, the matrix is removed through different solvents, while thermal methods such as pyrolysis or the fluidised-bed process utilise heat energy. Pyrolysis is the most widespread method, but the fibres reclaimed by this process are commonly contaminated with char residues from the matrix which can degrade the fibre properties [7].

After the recycling procedure, the majority of the rCF comes in a short and discontinuous form with the length of the fibres ranging from 0.1 mm to 60 mm [5,7].

One way to further process these short rCF could be the conversion into nonwovens. Nonwovens are a type of fabric in which the fibres are bonded by thermal, mechanical, or chemical treatments [8]. Since they do not require yarn, short fibres can be used. For rCF, wet-laying would be suitable for manufacturing the nonwoven mat. This method is derived from the paper-making process in which the fibres are dispersed in an aqueous solution. The fibre slurry is deposited on a wire screen to drain the liquid, leaving the fibres to form a web. This process is suitable for almost all fibres [9] and is more gentle towards the fibres in comparison to carding, especially for more brittle fibres such as carbon, which tend to break into shorter lengths during the carding operation [10]. In addition, the wet medium also provides a safer environment in the manufacturing process since rCF reclaimed by pyrolysis can contain very fine carbon particles that could contaminate the air and equipment [11,12].

The disadvantages for the wet-laying process are the relatively low thickness of the resulting rCF mats and the higher consolidation pressure needed to adjust the composite thickness and increase the fibre volume fraction [13,14].

In the effort to compensate for the brittle nature of the rCF, hybridisation with natural fibres such as flax can be considered. The term "hybridisation" refers to the combination of two or more fibres inside a matrix. Oftentimes, natural fibre and synthetic fibre are chosen in order to achieve a synergetic effect on the composite's properties [15]. Flax is one of the strongest natural fibres and is already used in a variety of composites [16,17]. With a tensile strength of 1.3 GPa and a Young's modulus of 54 GPa, its properties are comparable to glass fibres [18,19]. CF are superior in strength, with a tensile strength of 3–5 GPa, but they exhibit brittle behaviour because of their high modulus of 250–700 GPa [20]. In a hybrid, the CF would provide strength while flax balances the inherent brittleness of CF. CF/flax hybrids are already used for sporting goods because of their combined strength and damping properties since the damping coefficient of composites reinforced by woven flax can be up to four times higher than the ones reinforced by carbon only [16,21].

Much of the work reported in the literature about hybrid CF/flax composites [21–25] has been focused on the plies of woven fabric from continuous virgin CF and long flax fibres rather than nonwoven recycled short fibres as proposed in this work. Few studies have examined the wet-laying of short fibre composites with either flax fibres [26–30] or rCF [11–13,31,32]. Recently, Longana et al. [33] have investigated a rCF/flax hybrid manufactured through a water-assisted fibre alignment method that was successfully used with short rCF and rCF/glass fibre hybrids [34,35]. The blending of both fibre types could potentially decrease the tendency for catastrophic delamination as the mixing is more intimate, which can enhance the hybrid's properties [36].

In this research, short fibre rCF/flax composites were produced using wet-laying. To maximize the intermixing of the rCF/flax fibres for improved hybrid properties, the rCF, flax, and the poly-lactic acid (PLA) matrix fibres were dispersed together during the wet-laying process. The different fibre ratios were characterised through three-point-bending, scanning electron microscopy, and optical microscopy to optimise the processing parameters for future manufacturing.

2. Materials and Methods

2.1. Materials

Recycled CF and flax fibres were used as the reinforcement material, while the PLA in fibre form was the matrix constituent. CF reclaimed by pyrolysis, unsized, and chopped to the length of 12 mm with the diameter of 6–8 μm were purchased from ELG Carbon Fibre Ltd. (Coseley, UK). Scutched flax fibres in 10 mm lengths were supplied by FRD (Rosières-près-Troyes, France). For the matrix material, PLA slivers from Sirdar Spinning Ltd. (Wakefield, UK) were cut into 10 mm long segments.

2.2. Nonwoven Formation

Fibres of 2 g in weight were dispersed in 2 litres of water through a paper pulp disintegrator for 10,000 revolutions at a speed of 3000 revolutions per minute.

The dispersion was poured with additional water into a Handsheet Former manufactured by Mavis Engineering Ltd. (London, UK). This apparatus consisted of a base with a porous mesh that was connected to a drain pipe. A metallic cylinder closed the top which was filled with the fibre dispersion. Through draining the water, the fibres collected on the net at the bottom and formed a circular mat with a diameter of 16 cm. These were blotted with filter papers and pressed with a couch roll to extract excess water and smooth out the surface. Afterwards, the mats were left to dry by air convection for 24 h at room temperature.

Five different samples were made with the volume ratios of rCF and flax fibre as presented in Table 1.

Table 1. Fibre volume ratio of recycled carbon fibres (rCF) and flax for each sample.

Sample No.	1	2	3	4	5
rCF [%]	100	75	50	25	0
Flax [%]	0	25	50	75	100

For the dispersion of the fibres, all components of the mat had to be added in the desired proportions. To keep the grammage of the fabric constant at 100 g/m², 2 g of fibres were used per mat in the form of a disc with a 16 cm diameter. The target fibre volume fraction V_f for the final composite was chosen to be 30% since previous work [13,27,28,30] reported lowered mechanical properties at higher fibre fractions due to fibre breakage. Assuming $V_f + V_m = 1$ (V_m = matrix volume fraction), Equation (1) can be used to determine the corresponding weight composition of the fibres. One layer of sample 2 with 75% rCF and 25% flax is shown in Figure 1.

$$V_f = \frac{\rho_m W_f}{\rho_m W_f + \rho_f W_m}$$
(1)

where

ρ_f = Density of fibre [g/cm³]
ρ_m = Density of matrix [g/cm³]
W_f = Weight of fibres [g]
W_m = Weight of matrix [g]

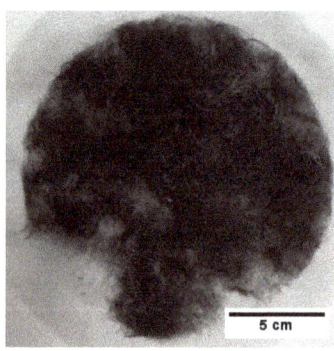

Figure 1. One layer of Sample 2 with 75% rCF and 25% flax.

Table 2 shows the fibre weights used for the mat formation of each sample calculated with $\rho_{rCF} = 1.8\ \text{g/cm}^3$, $\rho_{Flax} = 1.5\ \text{g/cm}^3$, and $\rho_{PLA} = 1.4\ \text{g/cm}^3$.

Table 2. Fibre weight compositions for one layer of each sample produced.

Sample No.	1	2	3	4	5
rCF [g]	0.71	0.54	0.37	0.19	0
Flax [g]	0	0.15	0.30	0.48	0.63
PLA [g]	1.29	1.31	1.33	1.33	1.37

2.3. Composite Fabrication

Several layers of the nonwoven mats with the same fibre ratio were stacked between two 0.2 mm thick PTFE (Polytetrafluorethylene) sheets before being placed inside the hot press. The PTFE prevented the composite from adhering to the steel plates of the machine. Upon closing the press, a 2 bar pressure was applied and the plates were heated at a rate of ca. 25 °C/min until a temperature of 160 °C was reached. This temperature was kept for 10 min while maintaining the pressure, after which the plates were cooled to 50 °C and the composite sample was removed. The densities of the composites were determined through the water immersion method and compared to their theoretical densities to calculate the void content. In Table 3, the number of layers used to obtain the target thickness of 2 mm along with the density values are presented.

Table 3. Number of layers and the thickness (mm) of the samples produced. Their measured and calculated densities along with the calculated void content are also displayed.

Sample No.	1	2	3	4	5
No. of layers	20	22	26	25	30
Thickness [mm]	1.95	1.98	2.09	1.95	2.03
Measured density [g/cm³]	1.07	1.33	1.39	1.32	1.32
Calculated density [g/cm³]	1.52	1.50	1.48	1.45	1.43
Calculated void content [%]	29.44	11.21	6.03	9.19	7.70

2.4. Optical Microscopy

Cut-outs in the dimension of $1 \times 2\ \text{cm}^2$ from each composite were cast in an epoxy resin and cured at room temperature for 24 h. The prepared specimens were ground and polished until the surface and the cross-section were sufficiently smooth to be viewed under edge-lighting of the Keyence VHX-5000 microscope. This technique allowed better distinguishing of the brighter reflective fibres and voids in the PLA matrix that appeared darker in the optical image.

2.5. Flexural Tests

Three-point bending tests were performed according to BS EN ISO 14125 to determine the flexural behaviour. Five specimens of each sample were tested on an Instron model 5969 with a 10 kN load cell at a cross-head speed of 1 mm/min. The deflection was given by the displacement of the central loading member, and the specimen dimensions were $60 \times 25\ \text{mm}^2$ with a span ratio of 1:20.

The flexural stress σ_f was determined by the following equation (Equation (2)):

$$\sigma_f = \frac{3FL}{2bh^2} \tag{2}$$

where

σ_f = flexural stress [MPa]
F = load [N]

L = length of span [mm]
b = width of the specimen [mm]
h = thickness of the specimen [mm]

Flexural modulus E_f was obtained by:

$$E_f = \frac{L^3 F}{4bh^3 d} \tag{3}$$

where

E_f = flexural modulus of elasticity [MPa]
d = central deflection [mm]

Flexural strain was calculated as:

$$\varepsilon = \frac{6dh}{L^2} \tag{4}$$

2.6. Scanning Electron Microscopy

After the flexural tests were performed, the specimens were viewed under the scanning electron microscope (SEM) in order to assess surface damage. The specimens were sputter coated with gold and examined with the Zeiss EVO60 at 8 kV.

3. Results and Discussion

3.1. Morphological Observations

The finished composite samples are depicted in Figure 2a–e. Figure 1 shows a single layer of the nonwoven mat. Although the single layers produced showed variations in areal density, by stacking multiple plies, the variations were evened out and ultimately formed a homogenous composite with a smooth surface.

Unlike flax fibres, rCF does not contain a collapsible lumen nor does it change its shape when compressed. Therefore, fewer plies were needed to achieve an overall thickness of 2 mm for the samples with higher rCF content. Another effect of the lower compactability of rCF was the tapered edge of the composites with high rCF content, prominently visible in Figure 2a, as the rCF spread out when consolidated in the hot press. Furthermore, cellulose-based fibres like flax form inter-fibre hydrogen bonds after drying from wet-laying, which contributed to the shape retention in the samples with higher flax fibre content, as seen in Figure 2b–e.

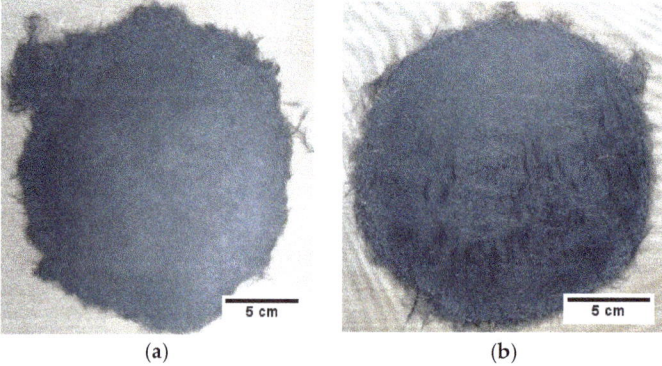

(a)   (b)

Figure 2. *Cont.*

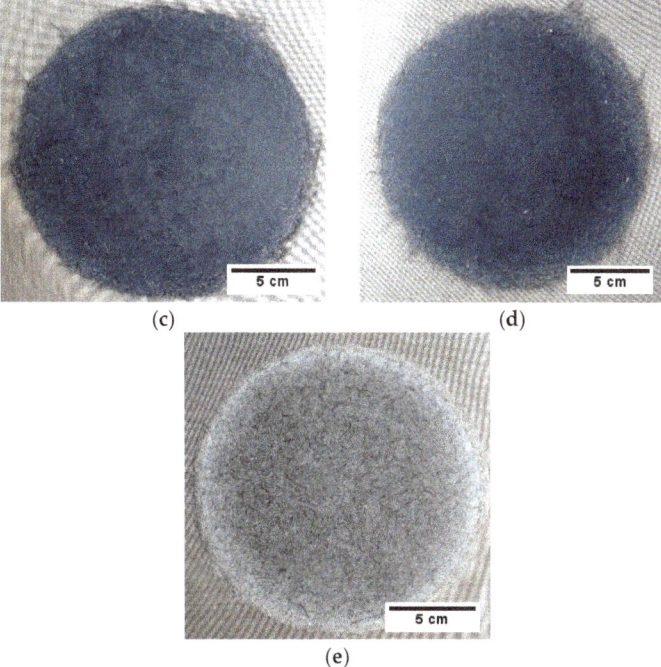

Figure 2. Finished composites of (**a**) sample 1, 100% rCF; (**b**) sample 2, 75% rCF and 25% flax; (**c**) sample 3, 50% rCF and 50% flax; (**d**) sample 4, 25% rCF and 75% flax; and (**e**) sample 5, 100% flax.

3.2. Analysis of the Cross-Section and Void Content

In Figure 3, the optical micrographs of the composites' cross-sections are presented. For sample 1 (Figure 3a), the view under the microscope revealed several voids inside the composite which explained the low density measured compared to the theoretical value. These almost round voids were in the size range of 10–100 μm and randomly distributed throughout the cross-section. The bright edges of the voids and the other bright spots were caused by the polished congregations of PLA which also reflected the light. This indicated that the 2 bar applied pressure was not enough to obtain sufficiently densely packed fibres for the target composite fibre volume fraction of 30% with a 100% rCF content.

Samples 2, 3, and 4 (Figure 3b–d) showed no visible voids and a fairly homogeneous distribution of rCF and flax fibres. The density measurements showed slightly elevated void contents for samples 2 and 4 at 9.1% and 11.2%, respectively, in comparison to samples 3 and 5 with 6.0% and 7.7%. They could be caused by increased gas formation from the PLA during the heating process on the hot press [37]. The higher void content from samples 2 and 4 were not detected through the microscope since the cross-section examined only represented a small part of the overall composite specimen and the voids might be present in other areas.

However, increasing the flax fibre content seemed to reduce the void content when compared to sample 1 by achieving the right amount of compactability for the PLA to fill in the space between the fibres and form a continuous solid phase. Replacing 25 vol % rCF with flax, e.g., sample 2 in Figure 3b, was enough to obtain the target fibre content without further increasing the pressure.

In sample 5, with 100% flax fibres (Figure 2e), thin voids were seen mostly surrounding tightly packed fibre bundles. This suggests that the fibre bundles were packed too closely for the PLA to wet the fibres. The elongated shape of the voids could potentially contribute to premature failure by initiating matrix cracking from the ends of the voids [19].

Generally, nonwoven composites will have higher void content compared to woven or unidirectional composites because the random orientation of the fibres can impede matrix flow or entrap air [38]. Furthermore, the hollow lumen inside natural fibres such as flax also act as a void and increase the void content of the composite [37].

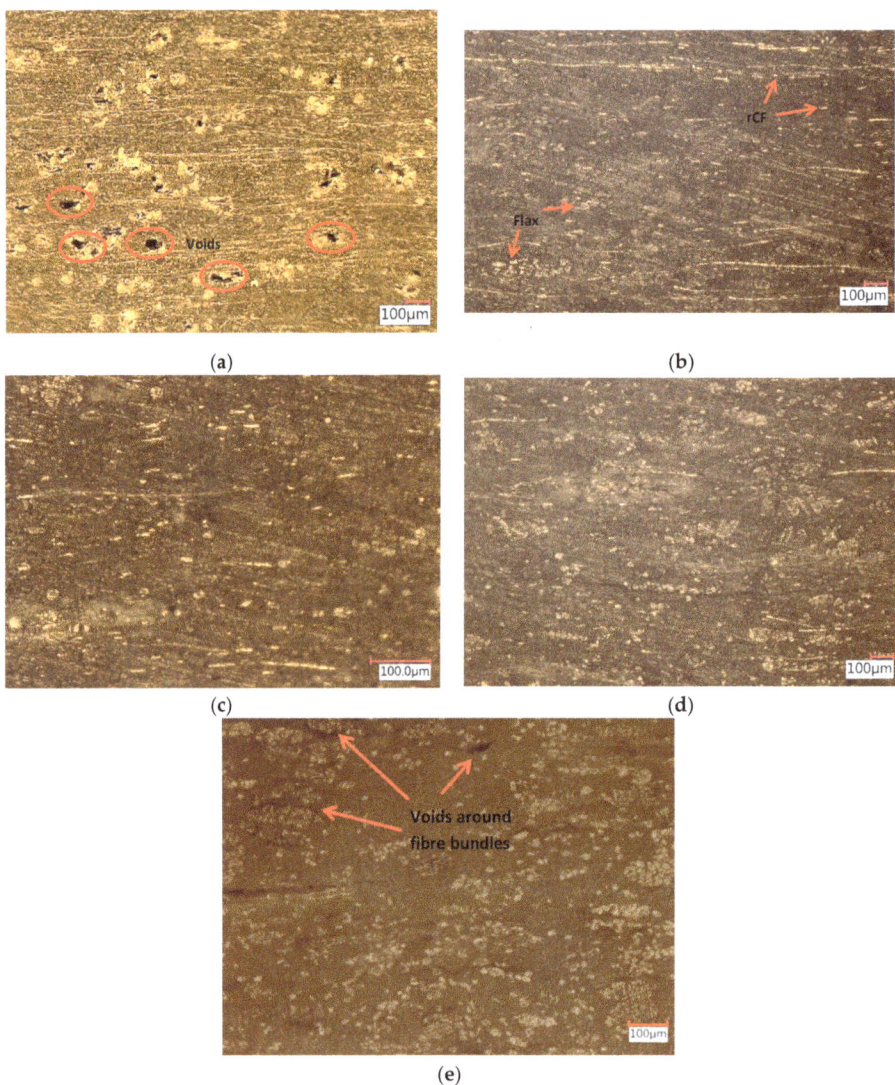

Figure 3. Optical micrographs with edge-lighting of the cross-section of (**a**) sample 1, 100% rCF; (**b**) sample 2, 75% rCF and 25% flax; (**c**) sample 3, 50% rCF and 50% flax; (**d**) sample 4, 25% rCF and 75% flax; and (**e**) sample 5, 100% flax.

3.3. Flexural Tests

The stress–strain curves obtained from the flexural test are plotted in Figure 4a–e. Table 4 presents the values for the flexural modulus, flexural strength, and strain to failure. The flexural modulus and flexural strength of the tested samples are also displayed in Figure 5.

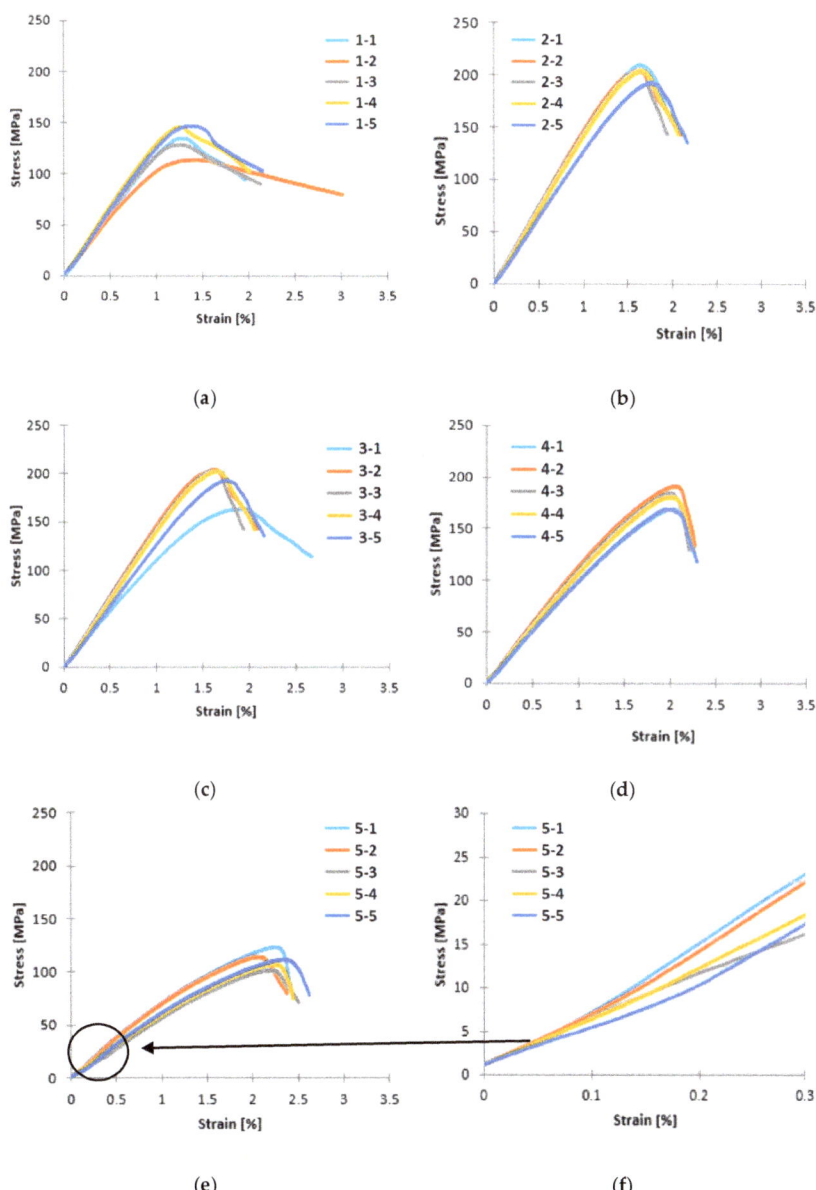

Figure 4. Flexural stress–strain curves for (**a**) sample 1, 100% rCF; (**b**) sample 2, 75% rCF and 25% flax; (**c**) sample 3, 50% rCF and 50% flax; (**d**) sample 4, 25% rCF and 75% flax; and (**e**) sample 5, 100% flax. (**f**) Crimp region of the stress–strain curve from sample 5.

Table 4. Flexural strength, flexural modulus, and strain at failure with standard deviations of all tested samples.

Sample No.	1	2	3	4	5
Flexural Modulus [GPa]	12.20	13.98	13.12	10.47	6.17
S.D.	0.90	0.97	1.03	0.73	1.17
Flexural Strength [MPa]	133.77	202.61	198.66	179.00	111.61
S.D.	13.70	5.9	23.21	9.84	8.40
Flexural Failure Strain [%]	2.26	2.08	2.04	2.26	2.47
S.D.	0.43	0.09	0.36	0.04	0.09

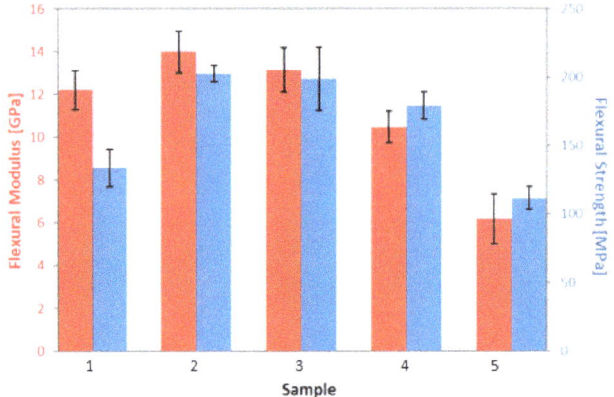

Figure 5. Flexural modulus and strength for all tested samples. Sample 1 showed lower properties due to voids in the matrix.

Below 0.3% strain, the stress–strain response showed a small crimp region for all curves. A magnification of the area up 0.3% strain for the stress–strain curve of sample 5 can be seen in Figure 4f. The crimp could be attributed to the elastic deformation of the PLA matrix. Its polymer strands were stretched and absorbed some of the stress as the cross-head touched the outermost layers at low strains. Afterwards, the curve became more linear with the increasing rCF content. The addition of flax fibres contributed to the non-linear deformation in the composite past the point of ultimate strength at which the matrix fails, and further softening of the curves could be seen for samples 1, 2, and 3 (Figure 4a–c) which corresponded to breakages of rCF. Overall, by increasing the flax content, the composite's ability to withstand higher strain was also increased due to the ductile behaviour of flax. Sample 1 seemed to break at a higher strain at failure on average because the high void content which introduced irregularities to the composite's composition, resulting in variable flexural behaviour. Although sample 3 contained more flax than sample 2, they behaved similarly and failed at around 2% strain. Sample 4 failed at 2.3% strain and sample 5 had the highest strain to failure at 2.5%, since this was made entirely of flax fibres.

In general, the composites with higher rCF content had higher flexural modulus and strength. For this reason, sample 1 with 100% rCF as reinforcement was expected to exhibit the highest flexural properties. However, its flexural modulus of 12.20 GPa was below those of samples 2 and 3 with 13.98 GPa and 13.12 GPa, respectively. Moreover, the flexural strength of sample 1 at 133.77 MPa was the second lowest after sample 5 with 111.61 MPa which only used flax fibres as reinforcement. This could be explained by the numerous voids observed in the optical micrographs of sample 1. Because of the high void fraction, the load could not be completely transferred to the rCF so that the matrix and the rCF broke separately from each other. This caused the softening after the point of ultimate strength in the stress–strain curve in Figure 4a.

It was found that sample 2 had the highest average flexural strength of 202.61 MPa followed closely by sample 3 with 198.66 MPa. Between samples 2 and 3, the increase of flax fibre content only decreased the flexural strength by 2.1% and the flexural modulus by 6.2%, which was within the scatter of the measured data. Further reducing the rCF content to 25% in sample 4 resulted in a decrease of almost 10% in flexural strength and 20% in flexural modulus when compared to sample 3. From the perspective of sample 5, replacing 25 vol % of flax with rCF increased the flexural strength by 60% and the flexural modulus by 70%. Compared to the relevant literature, Le Guen et al. [21] showed a decrease of 34.2% in flexural strength of their woven CF/flax epoxy composite as the CF volume fraction was lowered from 58% to 26%. However, this is a different system with long fibres and separate plies. This suggests that with the better intermixing of rCF/flax, the higher flexural properties could be maintained over a wider range of rCF/flax ratios; the void content and overall quality of the fabricated specimen are vital.

3.4. Surface Morphology and Failure Assessment

SEM results revealed that in addition to the voids inside sample 1, it also had a porous surface. The surface images of each sample are shown in Figure 6. Although the optical micrographs suggest an even mixture of the fibres, from the images it appeared as if the rCF in the composites tended to accumulate on the outermost layers and leave a rougher surface, as seen in Figure 6a–c. In contrast, samples with higher flax fibre content gave the surface a smoother appearance (Figure 6e).

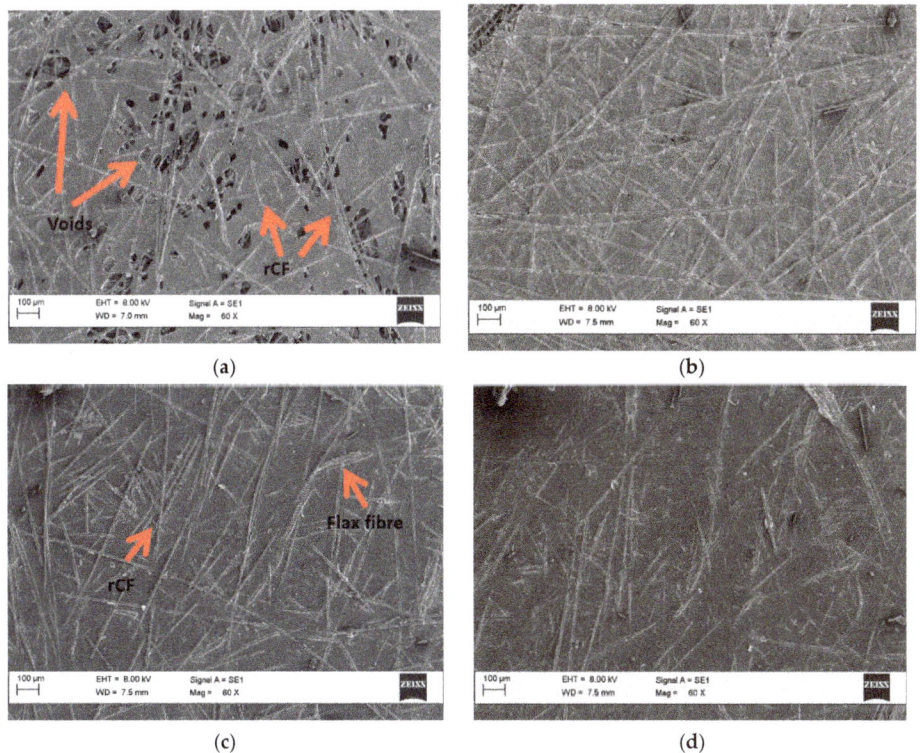

Figure 6. *Cont.*

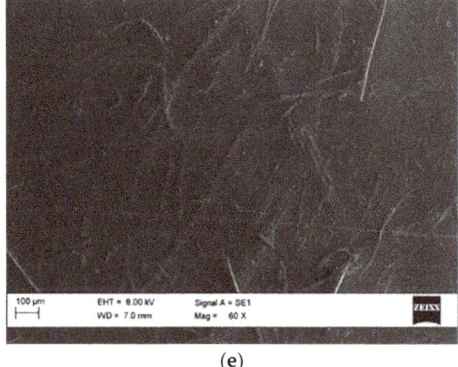

(**e**)

Figure 6. Scanning electron microscope (SEM) images at 60× magnification of the surface of (**a**) sample 1, 100% rCF; (**b**) sample 2, 75% rCF and 25% flax; (**c**) sample 3, 50% rCF and 50% flax; (**d**) sample 4, 25% rCF and 75% flax; and (**e**) sample 5, 100% flax.

The main cause of failure was the excessive tensile stress at the bottom surface of the specimen which resulted in the breakage of the matrix. This failure mode contributed towards the low flexural properties of sample 1 as well, since the propagation of the crack was facilitated by the presence of voids and surface porosity that magnified local stresses. Fibre pull-out and breakage were also visible, especially for the rCF, which could be seen in Figure 7. Most of the rCF fibres beneath the crack were not coated by matrix material, also visible in Figure 7, suggesting that they were de-bonded from the PLA matrix due to poor adhesion. This led to poor interfacial strength and hence a less effective load transfer from the resin to the fibres, resulting in premature failure and hence lower flexural strength. The matrix crack in sample 2 (Figure 8) was similar in size and shape to the one observed in sample 1. Fibre pull-out and breakages of rCF could be detected, but without the surface porosity, higher load was needed to propagate the crack through the matrix, which was why the average flexural strength of sample 2 was higher when compared to sample 1.

With higher flax content, more flax fibres were visible on the composite fracture surface. The flax fibres perpendicular to the direction of load could act as a crack initiator through de-bonding from the matrix, as seen in specimen 3-3 in Figure 9. Below the crack surface, flax fibres with matrix residue were visible which suggested better matrix adhesion. Specimen 4-3 in Figure 10 also had fibres on the surface; however, the fibres were not aligned in the direction of the crack and seemed to deflect it. It seemed as if the increase in flax fibre content decreased the rCF de-bonding from the matrix since the propagation of the crack appeared to be more winded and less prominent, e.g., in Figures 10 and 11.

From the higher magnification images of the composites with rCF, it can be concluded that the matrix adhesion between rCF and PLA tended to be poorer in comparison to flax as pull-outs and breakages of the rCF were clean. The higher affinity between flax and PLA is explained through their hydrophilic nature and the ability to form hydrogen bonds with their abundance of polar groups in their structures. Meanwhile, rCF containing mostly nonpolar C–C bonds was not able to form lasting hydrogen bonds with PLA and parted from the matrix with less force. The better matrix combability also resulted in a smaller matrix crack, as seen when comparing samples 1 and 2 with high rCF content and a gap of 100 μm in the matrix in (Figures 7 and 8) with sample 5 with 20 μm (Figure 11).

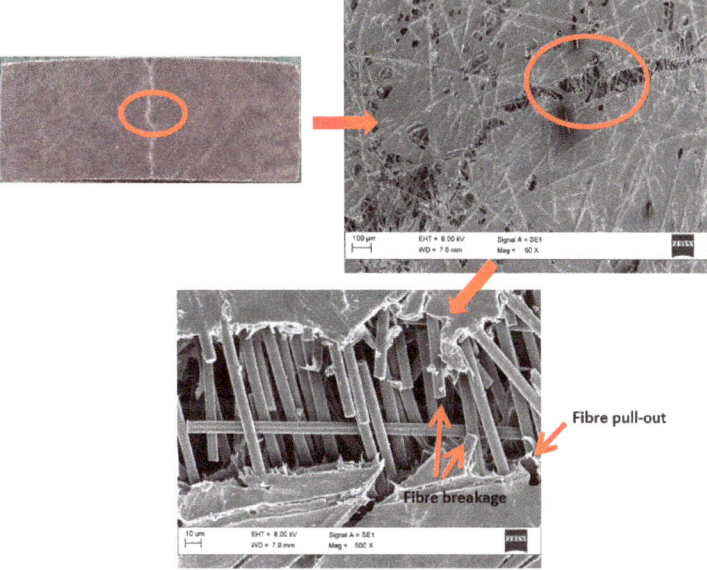

Figure 7. SEM images of the matrix crack of specimen 1-3 in 60× and 500× magnification. Fibre pull-out and breakages are visible.

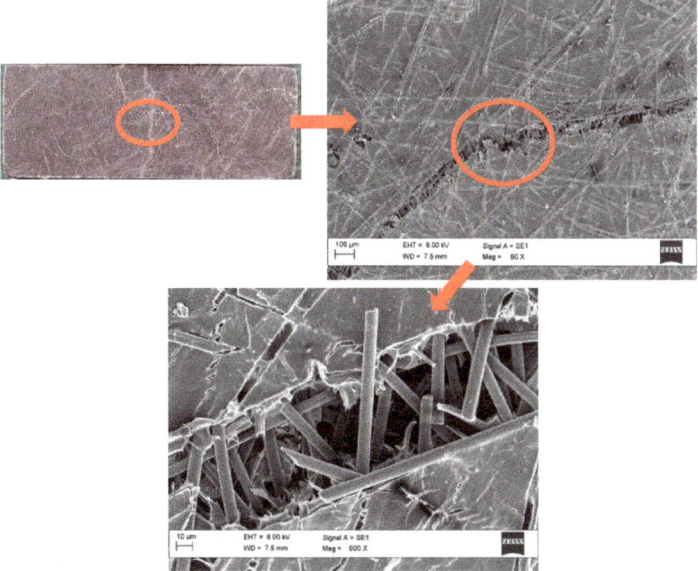

Figure 8. SEM images of the matrix crack of specimen 2-3 in 60× and 500× magnification.

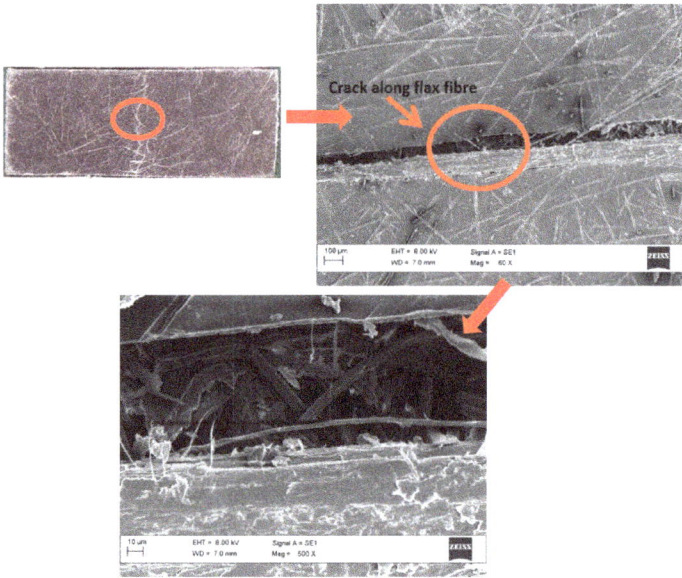

Figure 9. SEM images of the matrix crack of specimen 3-3 in 60× and 500× magnification. The crack propagated along the flax-fibre-PLA interface.

Figure 10. SEM images of the matrix crack of specimen 4-3 in 60× and 500× magnification.

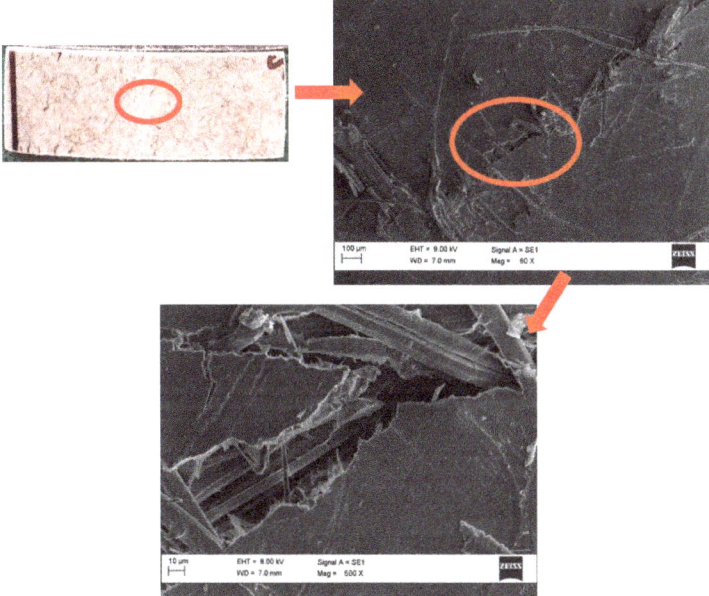

Figure 11. SEM images of the matrix crack of specimen 5-3 in 60× and 500× magnification.

4. Conclusions

Short fibre hybrid composites were manufactured through wet-laying with different ratios of rCF and flax fibres, and their flexural behaviour and morphological structure were analysed. The composite with 75% rCF and 25% flax fibre showed the highest average flexural strength and modulus of 202.61 MPa and 13.98 GPa, respectively, followed closely by the sample with 50% rCF and 50% flax. The intimate mixing between rCF and flax fibre during the dispersing stage allowed for a lesser decrease of flexural properties as the flax fibre content increased. Surface porosity and void content in the 100% rCF composite induced during the fabrication process contributed to early matrix cracking initiation that led to fibre/matrix de-bonding and premature fibre failure in the form of breakage and fibre pull-out. Replacing 25% of the rCF content with flax helped to increase the compactability of the nonwoven, resulting in the lowering of the void content without increasing the pressure of the hot press. Furthermore, rCF showed poor matrix adhesion with PLA, and fibre treatments could be considered to increase the interfacial bond between matrix and fibres.

Author Contributions: Conceptualization and methodology, B.T. and H.G.; validation, C.S. and H.G.; investigation, B.T. and X.Y.; writing–original draft preparation, B.T. and X.Y.; writing–review and editing, B.T., C.S. and H.G.; supervision, C.S. and H.G.

Funding: This project received funding from The European Union's Horizon 2020 research and innovation programme under grant agreement No. 690638.

Conflicts of Interest: The authors declare no conflict of interest.

References

1. Bunsell, A.R. Fibres for composite reinforcement: Properties and microstructures. In *Composite Reinforcements for Optimum Performance*; Woodhead Publishing: Sawston, UK, 2011; pp. 3–31, ISBN 9781845699659.
2. Kuehnel, M.; Kraus, T. The Global CFRP Market 2016. In Proceedings of the International Composites Congress (ICC), Düsseldorf, Germany, 28 November 2016.

3. Pimenta, S.; Pinho, S.T. Recycling carbon fibre reinforced polymers for structural applications: Technology review and market outlook. *Waste Manag.* **2011**, *31*, 378–392. [CrossRef] [PubMed]
4. Witik, R.A.; Teuscher, R.; Michaud, V.; Ludwig, C.; Månson, J.-A.E. Carbon fibre reinforced composite waste: An environmental assessment of recycling, energy recovery and landfilling. *Compos. Part A Appl. Sci. Manuf.* **2013**, *49*, 89–99. [CrossRef]
5. Pimenta, S.; Pinho, S.T. Recycling of Carbon Fibers. In *Handbook of Recycling: State-of-the-Art for Practitioners, Analysts, and Scientists*; Worrell, E., Reuter, M.A., Eds.; Elsevier Inc.: Amsterdam, The Netherlands, 2014; pp. 269–283. ISBN 9780123965066.
6. Pickering, S.J. Recycling technologies for thermoset composite materials—Current status. *Compos. Part A Appl. Sci. Manuf.* **2006**, *37*, 1206–1215. [CrossRef]
7. Oliveux, G.; Dandy, L.O.; Leeke, G.A. Current status of recycling of fibre reinforced polymers: Review of technologies, reuse and resulting properties. *Prog. Mater. Sci.* **2015**, *72*, 61–99. [CrossRef]
8. Russell, S.J. *Handbook of Nonwovens*; CRC Press: Boca Raton, FL, USA, 2006; ISBN 9781845691998.
9. Pill, H.; Afflerbach, K. Wet Lay Method. In *Nonwoven Fabrics*; Wiley-VCH Verlag GmbH & Co. KGaA: Weinheim, Germany, 2004; pp. 237–267.
10. Akonda, M.H.; Lawrence, C.A.; Weager, B.M. Recycled carbon fibre-reinforced polypropylene thermoplastic composites. *Compos. Part A Appl. Sci. Manuf.* **2012**, *43*, 79–86. [CrossRef]
11. Giannadakis, K.; Szpieg, M.; Varna, J. Mechanical Performance of a Recycled Carbon Fibre/PP Composite. *Exp. Mech.* **2011**, *51*, 767–777. [CrossRef]
12. Szpieg, M.; Wysocki, M.; Asp, L.E. Reuse of polymer materials and carbon fibres in novel engineering composite materials. *Plast. Rubber Compos.* **2009**, *38*, 419–425. [CrossRef]
13. Shah, D.U.; Schubel, P.J. On recycled carbon fibre composites manufactured through a liquid composite moulding process. *J. Reinf. Plast. Compos.* **2016**, *35*, 533–540. [CrossRef]
14. Turner, T.A.; Warrior, N.A.; Pickering, S.J. Development of high value moulding compounds from recycled carbon fibres. *Plast. Rubber Compos.* **2010**, *39*, 151–156. [CrossRef]
15. Nunna, S.; Chandra, P.R.; Shrivastava, S.; Jalan, A. A review on mechanical behavior of natural fiber based hybrid composites. *J. Reinf. Plast. Compos.* **2012**, *31*, 759–769. [CrossRef]
16. Pil, L.; Bensadoun, F.; Pariset, J.; Verpoest, I. Why are designers fascinated by flax and hemp fibre composites? *Compos. Part A Appl. Sci. Manuf.* **2016**, *83*, 193–205. [CrossRef]
17. Yan, L.; Chouw, N.; Jayaraman, K. Flax fibre and its composites—A review. *Compos. Part B Eng.* **2014**, *56*, 296–317. [CrossRef]
18. Baley, C. Analysis of the flax fibres tensile behaviour and analysis of the tensile stiffness increase. *Compos. Part A Appl. Sci. Manuf.* **2002**, *33*, 939–948. [CrossRef]
19. Hull, D.; Clyne, T.W. *An Introduction to Composite Materials*; Cambridge University Press: Cambridge, UK, 1996; ISBN 9781139170130.
20. Chung, D.D.L. *Carbon Fiber Composites*; Butterworth-Heinemann: Oxford, UK, 1994; ISBN 9780080500737.
21. Le Guen, M.J.; Newman, R.H.; Fernyhough, A.; Emms, G.W.; Staiger, M.P. The damping–modulus relationship in flax–carbon fibre hybrid composites. *Compos. Part B Eng.* **2016**, *89*, 27–33. [CrossRef]
22. Sarasini, F.; Tirillò, J.; D'Altilia, S.; Valente, T.; Santulli, C.; Touchard, F.; Chocinski-Arnault, L.; Mellier, D.; Lampani, L.; Gaudenzi, P. Damage tolerance of carbon/flax hybrid composites subjected to low velocity impact. *Compos. Part B Eng.* **2016**, *91*, 144–153. [CrossRef]
23. Dhakal, H.N.; Zhang, Z.Y.; Guthrie, R.; MacMullen, J.; Bennett, N. Development of flax/carbon fibre hybrid composites for enhanced properties. *Carbohydr. Polym.* **2013**, *96*, 1–8. [CrossRef] [PubMed]
24. Assarar, M.; Zouari, W.; Sabhi, H.; Ayad, R.; Berthelot, J.M. Evaluation of the damping of hybrid carbon-flax reinforced composites. *Compos. Struct.* **2015**, *132*, 148–154. [CrossRef]
25. Fiore, V.; Valenza, A.; Di Bella, G. Mechanical behavior of carbon/flax hybrid composites for structural applications. *J. Compos. Mater.* **2012**, *46*, 2089–2096. [CrossRef]
26. Bos, H.L.; Müssig, J.; van den Oever, M.J.A. Mechanical properties of short-flax-fibre reinforced compounds. *Compos. Part A Appl. Sci. Manuf.* **2006**, *37*, 1591–1604. [CrossRef]
27. Bodros, E.; Pillin, I.; Montrelay, N.; Baley, C. Could biopolymers reinforced by randomly scattered flax fibre be used in structural applications? *Compos. Sci. Technol.* **2007**, *67*, 462–470. [CrossRef]
28. Roussière, F.; Baley, C.; Godard, G.; Burr, D. Compressive and Tensile Behaviours of PLLA Matrix Composites Reinforced with Randomly Dispersed Flax Fibres. *Appl. Compos. Mater.* **2012**, *19*, 171–188. [CrossRef]

29. Fages, E.; Cano, M.; Gironés, S.; Boronat, T.; Fenollar, O.; Balart, R. The use of wet-laid techniques to obtain flax nonwovens with different thermoplastic binding fibers for technical insulation applications. *Text. Res. J.* **2013**, *83*, 426–437. [CrossRef]

30. Fages, E.; Gironés, S.; Sánchez-Nacher, L.; García-Sanoguera, D.; Balart, R. Use of wet-laid techniques to form flax-polypropylene nonwovens as base substrates for eco-friendly composites by using hot-press molding. *Polym. Compos.* **2012**, *33*, 253–261. [CrossRef]

31. Wong, K.H.; Syed Mohammed, D.; Pickering, S.J.; Brooks, R. Effect of coupling agents on reinforcing potential of recycled carbon fibre for polypropylene composite. *Compos. Sci. Technol.* **2012**, *72*, 835–844. [CrossRef]

32. Pickering, S.; Liu, Z.; Turner, T.; Wong, K. Applications for carbon fibre recovered from composites. *IOP Conf. Ser. Mater. Sci. Eng.* **2016**, *139*. [CrossRef]

33. Longana, M.L.; Yu, H.; Aryal, P.; Potter, K.D. The High Performance Discontinuous Fibre (HiPerDiF) Method for Carbon-Flax Hybrid Composites Manufacturing. In Proceedings of the 21st International Conference on Composite Materials, Xi'an, China, 20–25 August 2017.

34. Longana, M.L.; Yu, H.; Potter, K.D. The High Performance Discontinuous Fibre (HiPerDif) Method for the Remanufacturing of Mixed Length Reclaimed Carbon Fibres. In Proceedings of the 21st International Conference on Composite Materials, Xi'an, China, 20–25 August 2017.

35. Yu, H.; Potter, K.D.; Wisnom, M.R. A novel manufacturing method for aligned discontinuous fibre composites (High Performance-Discontinuous Fibre method). *Compos. Part A Appl. Sci. Manuf.* **2014**, *65*, 175–185. [CrossRef]

36. Flynn, J.; Amiri, A.; Ulven, C. Hybridized carbon and flax fiber composites for tailored performance. *Mater. Des.* **2016**, *102*, 21–29. [CrossRef]

37. Alimuzzaman, S.; Gong, R.H.; Akonda, M. Impact Property of PLA/Flax Nonwoven Biocomposite. *Conf. Pap. Mater. Sci.* **2013**, *2013*, 136861. [CrossRef]

38. Yahaya, R.; Sapuan, S.M.; Jawaid, M.; Leman, Z.; Zainudin, E.S. Effect of fibre orientations on the mechanical properties of kenaf–aramid hybrid composites for spall-liner application. *Def. Technol.* **2016**, *12*, 52–58. [CrossRef]

Article

Flexural Mechanical Properties of Hybrid Epoxy Composites Reinforced with Nonwoven Made of Flax Fibres and Recycled Carbon Fibres

Jens Bachmann *, Martin Wiedemann and Peter Wierach

DLR—Deutsches Zentrum für Luft-und Raumfahrt e.V. (German Aerospace Centre),
Institute of Composite Structures and Adaptive Systems, Braunschweig 38108, Germany;
martin.wiedemann@dlr.de (M.W.); peter.wierach@dlr.de (P.W.)
* Correspondence: jens.bachmann@dlr.de; Tel.: +49-531-295-3218

Received: 3 September 2018; Accepted: 3 October 2018; Published: 10 October 2018

Abstract: Can a hybrid composite made of recycled carbon fibres and natural fibres improve the flexural mechanical properties of epoxy composites compared to pure natural fibre reinforced polymers (NFRP)? Growing environmental concerns have led to an increased interest in the application of bio-based materials such as natural fibres in composites. Despite their good specific properties based on their low fibre density, the application of NFRP in load bearing applications such as aviation secondary structures is still limited. Low strength NFRP, compared to composites such as carbon fibre reinforced polymers (CFRP), have significant drawbacks. At the same time, the constantly growing demand for CFRP in aviation and other transport sectors inevitably leads to an increasing amount of waste from manufacturing processes and end-of-life products. Recovering valuable carbon fibres by means of recycling and their corresponding re-application is an important task. However, such recycled carbon fibres (rCF) are usually available in a deteriorated (downcycled) form compared to virgin carbon fibres (vCF), which is limiting their use for high performance applications. Therefore, in this study the combination of natural fibres and rCF in a hybrid composite was assessed for the effect on flexural mechanical properties. Monolithic laminates made of hybrid nonwoven containing flax fibres and recycled carbon fibres were manufactured with a fibre volume fraction of 30% and compared to references with pure flax and rCF reinforcement. Three-point bending tests show a potential increase in flexural mechanical properties by combining rCF and flax fibre in a hybrid nonwoven.

Keywords: composite; natural fibre; flax; recycled carbon fibre; nonwoven; hybrid

1. Introduction

Fibre reinforced polymers (FRP) have gained importance in aviation and other transportation sectors due to their excellent mechanical properties combined with relatively low weight. High performance composites like carbon fibre reinforced polymers (CFRP) and also glass fibre reinforced polymers (GFRP) are used in primary and secondary structures of modern aircrafts. They enable the construction of lighter and more efficient aircraft resulting in the reduction of fuel consumption and increased payloads Carbon fibres consume high amounts of energy during the production phase. Therefore, it is of high interest to reduce the consumption of synthetic materials in favour of bio-based materials in certain applications. Bio-based (renewable) materials like natural fibres have been under investigation for a long time for their use in composites but they have not yet been introduced into modern aircraft in a noticeable way. Lack of experience and confidence in the long-term performance and mechanical properties of composites containing natural fibres are still an obstacle for their usage in safety relevant applications like primary structures (e.g., fuselage). However, secondary structures and interior composites, which are not stressed on such high levels offer possible areas of application in aviation [1].

In contrast to synthetic fibres, natural fibres are characterised by a complex multiscale structure, leading to a large variability in mechanical properties for different natural fibres [2]. Compared to glass fibres, natural fibres usually offer good specific stiffness due to their low density. However, their tensile strength cannot compete even when taking into account the fibre density. Furthermore, the length of natural fibres is limited to the maximum of the plant length. Single flax fibres reach a maximum length in the two-digit millimetre range. This is a major difference compared to synthetic fibres which are available as filaments in theoretically unrestricted length. A comprehensive review of natural fibres and their properties can be found in the literature [3–7].

The mechanical properties of natural fibre reinforced polymers (NFRP) are typically lower when compared to GFRP and especially CFRP. In order to broaden the application of NFRP, it is important to increase their mechanical properties. Several ways to improve the mechanical properties of NFRP have been investigated [8]. Most of them use chemical treatments to improve the fibre-matrix adhesion. For example, the effect of silane coupling agents on NFRP has been reviewed by Xie et al. [9]. As another example, the positive effects of the grafting of flax fibres with nanoparticles and incorporation of carbon nanotubes on natural fibres was recently reviewed by Li et al. [10].

Another way to increase the mechanical properties of NFRP is the hybridization with synthetic fibres such as glass or carbon fibres. There is a differentiation between interlayer (interply) and intralayer (intraply) hybrid composites [11,12]. A common configuration of hybrid composites is the interlayer (Figure 1a) because it is simple to produce by stacking commercially available reinforcement layers with different types of fibres, such as carbon fibre and glass fibre woven fabrics. A mix of different fibre types in one layer characterises intralayer hybrid composites, resulting in a higher dispersion of fibres but also a more complex production process (Figure 1b). Swolfs et al. [11] reported the limited availability of investigations in the comparison of interlayer and intralayer hybrid composites. Smaller delamination areas have been found by Park et al. [13] after impact tests in intralayer compared to interlayer hybrids aramid/polypropylene fibre composites. An increased resistance to crack propagation has been found by Pegoretti et al. [14] for E-glass and polyvinyl alcohol woven fabrics.

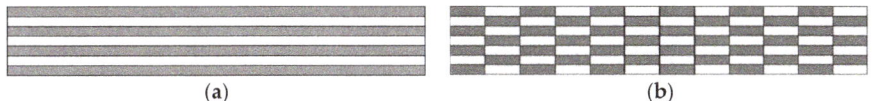

| (a) | (b) |

Figure 1. Hybrid configurations: interlayer (**a**) and intralayer (**b**). Dark and light grey fields indicate different types of fibres in the hybrid laminate. Based on [11].

In the literature, a multitude of hybrid combinations of all kind of natural fibres with glass, aramid and carbon fibres can be found. Manders and Bader [15] evaluated the tensile mechanical properties of glass/carbon fibre hybrid composites with epoxy resin. They found that the dispersion and localisation of the fibres had a strong effect on the results. An increase in toughness and failure strain was observed with a finer dispersion ("hybrid effect"). Reddy et al. [16] found a gradual increase in tensile and flexural modulus by stacking layers of jute, pineapple leaf and glass fibres. Lützkendorf et al. [17] assessed the combination of recycled carbon fibres (rCF) and natural fibres in a hybrid nonwoven and thermoplastic polypropylene matrix. With bonding agent, the flexural stiffness could be increased by more than 100% with an rCF-content of 20%. Using rCF/PP only in the top-layers with pure NF/PP in between (interlayer hybrid) led to a minor increase in flexural stiffness. Flynn et al. [18] tested the effects of hybridization with flax and carbon fibre fabrics in different stacking configurations. A gain in tensile strength of 252% compared to a purely flax fibre reinforced composite was found for the hybrid variant. Adekunle et al. [19] tested different hybrid combinations of woven and nonwoven flax fibres with glass fibres using a soybean oil derived bio-resin. They found a considerable increase in tensile strength by integration of a glass fibre mat. Cicala et al. [20] combined flax and carbon fibres in different stacking sequences and found up to three times higher tensile strength compared to the pure flax composite. Murdani et al. [21] found that the damping ratio decreased while flexural properties were improved by adding glass and carbon fibres to a jute fibre reinforced composite.

Summarizing the available literature on hybrid natural fibre/synthetic fibre composites, an improvement in mechanical properties was usually observed. A majority of studies are based on reinforcements using usual (commercial) delivery forms of reinforcement, such as woven fabric. However, the increasing use of CFRP in aviation and other high performance applications leads to a further rise in carbon fibre production. Double-digit growth for the demand of carbon fibre is expected in the next decade [22]. From an ecological perspective, composites like CFRP consume high amounts of energy during the production phase. The high value of carbon fibres makes them very interesting for recycling. Recycled carbon fibres can be obtained from production waste (dry fibres from cutting and trimming as well as uncured prepreg) and cured end-of-life CFRP products. Close-loop recycling, as with metals is currently not available for end-of-life CFRP because of their intrinsic heterogenic structure. Especially in combination with cross-linked thermoset polymer systems, this impedes efficient recycling at the end of life [23].

Potential cost savings and reduced energy consumption through carbon fibre recycling were predicted by Carberry [24]. However, recycling processes can be even more expensive compared to the production of vCF [25]. A comprehensive overview of composite recycling processes can be found in the review papers of Oliveux et al. [26] and Naqvi et al. [27]. Currently, the CFRP recycling process with the highest technology readiness level is pyrolysis [28]. CFRP waste is fed into the pyrolysis chamber at temperatures up to 600 °C. Pyrolysis has already found its way into commercial applications in the UK and Germany [29]. Typically, the quality of the recycled carbon fibres (rCF) is lower compared to virgin carbon fibres (vCF). A main reason is the discontinuous length due to the preceding shredding process to reduce the size of end-of-life parts. The removal of fibre sizing and thus a possible reduction of fibre-matrix adhesion is another drawback. Information on the influence of the fibre-matrix interface and interphase on the composite properties can be found in the work of Jesson et al. [30]. The mechanical properties of rCF can still be considered as very good. Tensile fibre tests have shown comparable results for rCF and vCF [31]. Because of the reduced length, the application of rCF is currently restricted to alternative use-cases with lower quality requirements [28]. The usual delivery forms are chopped, milled and pelletized rCF [29]. Another way to obtain recycled fibres is the recycling of waste of carbon fibre fabrics generated during the production of composites. Waste from fabric cutting is the main source for this material. The short and variable length of rCF is a challenge for converting them into continuous yarns used in woven reinforcements. Nonwovens made of chopped rCF are already available on the market [32,33].

Chopped rCF (Figure 2b) shows some apparent similarities to natural fibres (Figure 2a), for example, randomness, restricted length and curvature. Therefore, a possible way of using rCF is in combination with natural fibres in a hybrid nonwoven as reinforcement for FRP, combining the very good mechanical properties of the rCF with the low density and good damping properties of natural fibres. Nonwoven processes are capable of combining different types of fibres of variable length in a single web structure. Nonwovens are also less expensive and potentially more eco efficient compared to classic woven fabrics from bio-fibres due to their simple production process [34]. The question is, do the added steps in the production of a hybrid nonwoven add benefits compared to a more common reinforcement, which uses just one fibre type in woven or nonwoven fabrics? The aim of this study is to make a preliminary assessment of the impact of flexural mechanical properties. An overview of typical manufacturing processes for nonwoven fabric in relation to fibre length and degree of isotropic behaviour is given in Figure 3. A distinction can be made between four processes, and the two dry-laying processes are capable of processing fibre lengths of more than 30 mm. Aerodynamic airlay processes usually utilize an airstream to feed the fibres on a moving belt. Carding is a process of separating individual fibres, using a series of dividing and re-dividing steps. This results in a parallelisation of the fibres. Carding is possible by using hand cards or drum cards in which fibres are fed through one or more pinned drums. The nonwoven web can be parallel or random laid. Parallel laid carded nonwoven usually results in good tensile strength and low elongation in the machine direction (MD) compared to the cross direction (CD) [34].

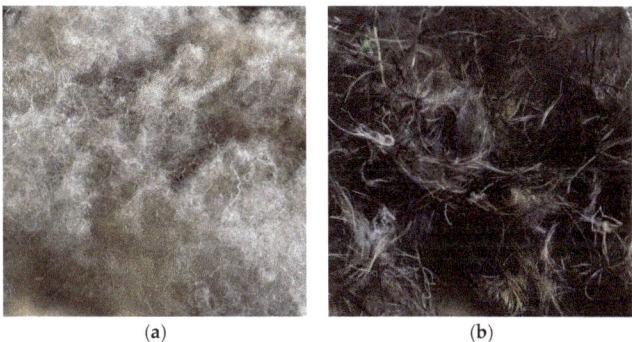

(a)   (b)

Figure 2. The delivery form of the fibres used for the assessment of hybrid nonwoven in this study. Flax fibres (**a**) and recycled carbon fibres (**b**).

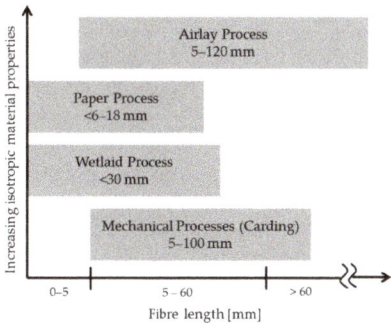

Figure 3. Manufacturing processes for nonwoven in relation to fibre length and degree of isotropic material properties. Figure based on [35].

The manufacturing processes of nonwoven fabrics are very efficient and it is possible to use different fibre length fractions. For woven fabrics, the fibres are normally processed into yarns, which leads to strong twists and thus to areas of potentially poor wettability by liquid resin systems. A possible advantage of the use of nonwoven for FRP lies in the better availability of the fibres to be embedded in the resin system. Nonwoven is an obvious choice because of the fibre characteristics of the flax and rCF. However, fibres can be damaged because of the mechanical stress during the carding process [35].

The hybrid combination of recycled carbon fibres with natural fibres and their different distribution in thermoset composites has not been assessed in detail so far. This study aimed to provide a preliminary assessment of the potential of hybrid rCF/flax nonwoven as reinforcement in combination with a thermoset resin. Two references, pure flax and pure rCF nonwoven reinforced composite, were produced as a base line. Two hybrid variants contained fibres in the same total volumetric mixing ratio of flax to rCF (3:1). The difference was the distribution of the rCF over the laminate thickness. The effect on flexural mechanical properties was studied by three-point bending (3PB). This test was chosen because of the simple specimen preparation and the small size of the samples.

2. Materials and Methods

2.1. Materials

The four variants of laminates tested in this paper were based on recycled carbon fibres from production waste (e.g., dry fibre cut-off) and staple flax fibres. Flax fibres were obtained from the company Intercot in Barcelona, Spain. The flax fibres had no length restriction due to chopping. To avoid the potentially weak fibre-matrix adhesion of pyrolysed rCF because of their removed sizing, recycled fibres from dry cut-off waste were chosen for this preliminary study. Those rCF usually have an intact sizing. However, the exact type of sizing was unknown. These rCF, chopped to maximum length of 25 mm, were obtained from CarboNXT GmbH (Wischhafen, Germany). The delivery form of the fibres is shown in Figure 2. A thermoset matrix system, the two-component liquid epoxy infusion resin Epikote™ RIMR135 with curing agent RIMH1366 (Hexion B.V., Rotterdam, The Netherlands) was used to produce the laminates.

2.2. Manufacturing of the Nonwoven

Sensitivity of electrical equipment to short circuits caused by conductive rCF led to the decision to perform the whole nonwoven manufacturing process in a closed exhaust hood. Therefore, a simple two-stage laboratory process was used to manufacture the nonwoven from flax and recycled carbon fibres. The first step in the process was opening of the fibres. The resulting increased volume of the fibre batches led to better control of the fibre feed into the card. The fibre opening process used in this study was based on the principle of a gas jet mixer. A container with a volume of approximately seven litres was modified for fibre opening and mixing, as shown in Figure 4a. Compressed air (2–3 bars) was introduced to the fibre filled container through a funnel attached the bottom. A manually controlled pistol was used to control the release of the compressed air in a pulsating manner. Circular distributed air outlets with cotton wool filter elements were installed in the upper part to the capsule. The outflowing compressed air took up the fibres and whirled them around in the capsule, resulting in a fibre opening (separation). This is schematically shown in Figure 4a. Opening and mixing processes also removed shives (wooden residues of the bast fibre stem) and dust which was gathered in the cotton wool filters. A transparent screw cap was used to visually observe the opening process. The fibre mixing was carried out in the same modified container that was used for the fibre opening. After opening, fibre batches were mixed according to the mixing ratio given in Table 1.

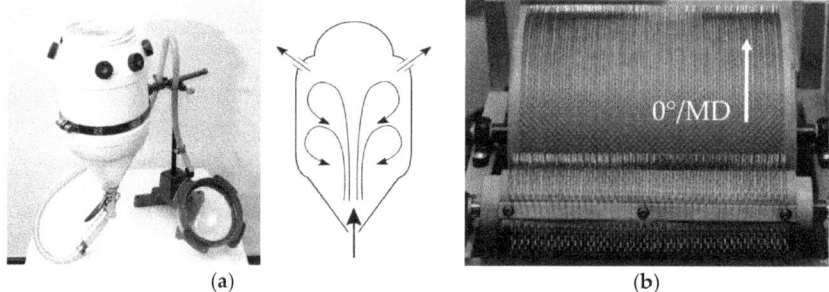

(a)   (b)

Figure 4. Reinforcement manufacturing process with fibre opening and mixing in a capsule with turbulent air (**a**) and nonwoven web formation on a small-scale electric card with two rotating drums (**b**).

Table 1. Nomenclature, composition and schematic drawings of the volumetric distribution of flax (light grey) and rCF (dark grey). Total fibre volume content for each composite was approximately 30%. The volumetric rCF to flax ratio in the outer layers of the Gr-22.5Flax-7.5rCF laminate was rCF/flax = 3:1 (intralayer). The four inner layers contained 100% flax fibres, resulting in a combination of intralayer and interlayer hybrid. The same total amount of flax and rCF was used for the 22.5Flax-7.5rCF laminate (pure intralayer).

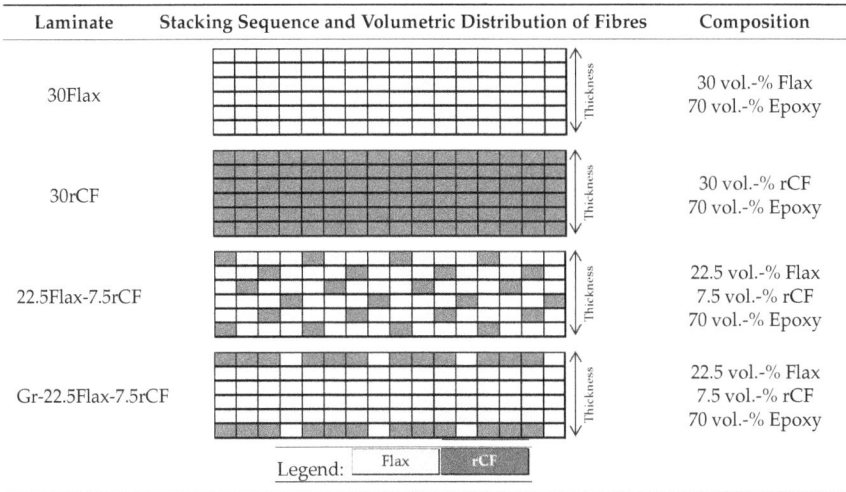

Laminate	Stacking Sequence and Volumetric Distribution of Fibres	Composition
30Flax		30 vol.-% Flax 70 vol.-% Epoxy
30rCF		30 vol.-% rCF 70 vol.-% Epoxy
22.5Flax-7.5rCF		22.5 vol.-% Flax 7.5 vol.-% rCF 70 vol.-% Epoxy
Gr-22.5Flax-7.5rCF		22.5 vol.-% Flax 7.5 vol.-% rCF 70 vol.-% Epoxy

Legend: Flax rCF

For the nonwoven web formation, an electric drum card with a simple two drum design was purchased from the company Holzwurmwolle in Alfdorf, Germany) (Figure 4b). Both drums had a width of 190 mm. The diameter of the small drum was 45 mm. The big drum had a diameter of 160 mm, resulting in a circumferential length of approximately 500 mm. Both drums could be individually activated and adjusted for rotating direction and rotating speed. Carding cloths with buckled needles of 17 mm length were attached to the drums. Needle density was 52 ppsi for the small intake drum and 72 ppsi for the big drum. Nonwoven webs with an area of approximately 500 mm length to 190 mm width were produced.

The opened fibres were fed manually to the electric drum carding device. A possible result of the manual feeding process was the increased risk of uneven fibre distribution (areal weight) in the nonwoven. This effect was reduced by quality control and the stacking process of several nonwoven layers to even out the distribution of fibres in the laminate. The fibres fed manually to the carding device were taken up by the small entrance roll and fed to the counter rotating big drum. This process step was carried out until all fibres were taken up from the big drum. Compression of the fibres on the big drum was carried out with a fixed brush. A flat card with a needle felt of 52 ppsi pressed manually on the big drum during the fibre take-up process led to a further increased fibre density of the nonwoven web. The last step was cutting and manual removal of the nonwoven from the big roll followed by quality control. No further compaction was applied to the nonwoven webs before laminate manufacturing. The quality control included the measurement of areal weight and checking the distribution of fibres with the help of a transmission light desk.

2.3. Composition and Nomenclature of the Composites

The stacking sequence and number of nonwovens used for the composite manufacturing is shown in Table 1. Four variations of laminates with a total of six nonwoven layers were produced. All layers were stacked symmetrically to the neutral axis and oriented in the same direction. The machine direction (MD) of the nonwoven was always oriented in 0° laminate direction. As the carding process led to a clear orientation of fibres in MD compared to the cross direction (CD), the laminates had no in-plane isotropic characteristics.

The nomenclature used for the laminates in this paper is a combination of fibre volume fraction in percent and the used fibre, e.g., 30Flax = 30% fibre volume fraction of flax fibres. Two non-hybrid reference laminates contain nonwoven with pure flax (30Flax) and pure rCF (30rC) reinforcements. Furthermore, two hybrid variants with an rCF to flax ratio of 1:3 but differing fibre distribution were manufactured. Of these, the intralayer hybrid variant contained 22.5 vol.-% flax fibres and 7.5 vol.-% rCF (=22.5Flax-7.5rCF) which were mixed with the aim of achieving a homogeneous distribution. The gradient or sandwich like variant was a combination of interlayer and intralayer hybrid nonwoven (Gr-22.5Flax-7.5rCF). Here, the rCF were distributed only in each outer layer mixed with a smaller amount of flax fibres added (intralayer). In these outer layers, the nonwoven had a flax rCF to flax ratio of 3:1. The four layers in between were purely made of flax fibres. The gradient variant was chosen in order to assess the effect of the concentration of rCF on the flexural mechanical properties. Placing stiffer plies (i.e., layers with higher rCF content) away from the neutral axis should result in increased flexural mechanical properties [12]. For both hybrid variants the same total mixing ratio of rCF and flax fibres was applied.

2.4. Manufacturing of the Composites

The two-component epoxy resin RIMR135 and hardener RIMH1366 were mixed in a weight ratio of 100:35, followed by a degassing step in a desiccator to remove air bubbles. The nonwoven layers have been stacked in order to achieve the desired fibre volume fraction and fibre distribution at an intended thickness of 3 mm. The stacking sequence, number of layers and nomenclature of the laminates is shown in Table 1. The monolithic laminates were produced with the single line infusion (SLI) method in a Lauffer hydraulic press (500 mm × 500 mm pressing area). The SLI process is characterised by using the same line for vacuum generation followed by the liquid resin infusion. Curing time in the heated hydraulic press was 120 min at 85 °C followed by deforming and post-curing at 100 °C for 60 min in a Memmert UFP500 convection oven (Memmert GmbH + Co. KG, Schwabach, Germany). After cutting, ultrasonic testing in water were carried out to assess the laminate quality regarding pore distribution and delamination. The cured and trimmed laminates were stored at 23 °C and 50% relative humidity. The physical properties of the laminates are summarized in Table 2.

Table 2. Measured and calculated physical properties of the cured monolithic composites.

Laminate	Fibre Volume Content [1] [%]	Average Thickness [2] [mm]	Density [3] [g/cm³]	Void Content [4] [%]	Glass Transition Temp. [5] [°C]	Water Content [6] [%]
30Flax	29.1	3.11	1.16	7.7	82.3	1.94
30rCF	29.0	3.16	1.29	4.3	87.5	0.46
22.5Flax-7.5rCF	29.5	3.12	1.24	2.9	85.2	1.65
Gr-22.5Flax-7.5rCF	29.9	3.07	1.23	3.8	85.8	1.45

[1] Calculated from average measured laminate thickness given in this table and nonwoven areal weight used for the manufacturing of the composites, with a fibre density of 1.5 g/cm³ for flax, 1.78 g/cm³ for the rCF and 1.15 g/cm³ for the epoxy resin; [2] Measured with a micrometer at nine points evenly distributed on the composite plates. The maximum standard deviation was 0.07 mm; [3] Calculated from measured laminate dimensions and weight after storing at 23 °C and 50% r.h. for at least 48 h; [4] Calculated from the difference of theoretical density and measured density; [5] T_G 2% (onset) measured with DMA (Dynamic Mechanical Analysis) on Mettler Toledo DMA SDTA861ᵉ. Three specimens with a size of 80 mm by 5 mm have been tested in 0° laminate direction. The maximum standard deviation was calculated to be 2.1 °C; [6] Measured with TGA (Thermogravimetric Analysis) on Mettler Toledo TGA/DSC3⁺ Starᵉ System (1st RT-120 °C, 2 K/min, N2; 2nd 120 °C, 180 min, N2) with a standard deviation of 0.02% for two samples.

Figure 5 gives an overview of the obtained distribution of the rCF in the laminates 30rCF (a), 22.5Flax-7.5rCF (b) and Gr-22.5Flax-7.5rCF (c). The microscopic images were processed in such a way that the rCF appear as white pixels while all other constituents (epoxy resin, flax fibres) and voids appear as pure black pixels. However, the exact fibre volume content could not be calculated by measuring the total area of white pixels. The rCF were located in a different angle to the image plain, resulting in ellipses instead of circles. This indicates the randomness of the fibre orientation typically seen in nonwoven fabric. In Figure 5a, the higher amount 30 vol.-% rCF compared to 7.5 vol.-% rCF in Figure 5b,c is obvious. For all laminates, the distribution of rCF and flax fibres was not homogeneous. This can be seen in the areal concentration of white pixels. Furthermore, Figure 5c shows the concentration of rCF in both outer layers of the laminate. Undulations and varying thicknesses of the rCF rich layers can be observed and are typical for all produced samples.

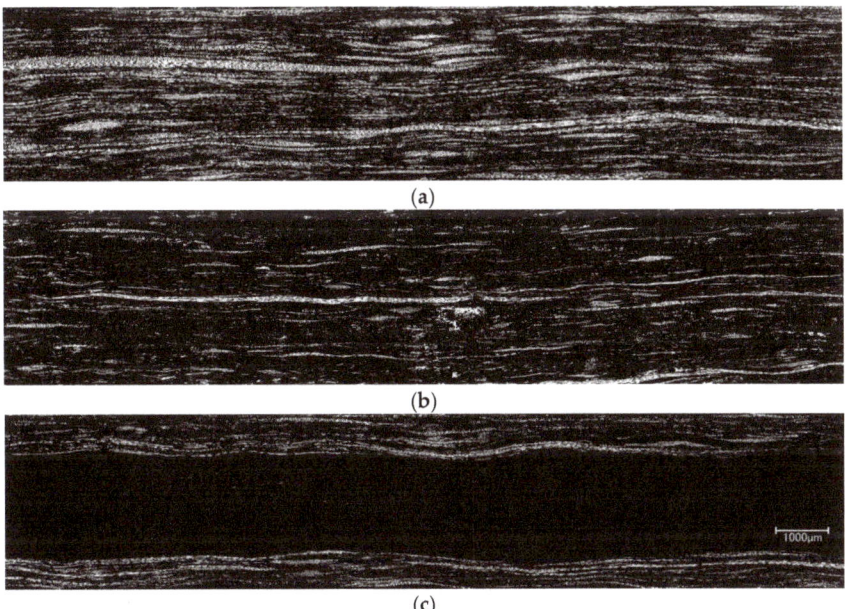

(a)

(b)

(c)

Figure 5. Visualisation of the typical rCF distribution shown as white areas. For the laminates 30rCF (**a**) and 22.5Flax-7.5rCF (**b**), the rCF should have been evenly distributed all over the composite. As not all fibres where fully opened, fibre rich areas as well as resin rich areas occurred. The laminate Gr-22.5Flax-7.5rCF (**c**) contains the same volumetric amount of rCF compared to (**b**), but is concentrated in both outer nonwoven layers. Micrographs were taken with Keyence VFX1000 and VH-Z100UR lens (×200) and stitched to a horizontal panorama. The image plane is the 90° laminate direction. All images were converted to black (flax, epoxy resin, voids) and white pixels (rCF) with the colour threshold tool in ImageJ software.

More detailed microscopic images of all four tested variants are shown in Figure 6. Here, the flax fibres and rCF are clearly visible. Flax fibres occur both as single fibres as well as in accumulations of single flax fibres (technical fibres). Voids can be observed especially around these flax fibre accumulations (Figure 6a). Flax fibres oriented perpendicular or in an angle between 0° and 90° appear as elongated dark voids (Figure 6c), most likely resulting from the removal of flax fibre segments during the polishing process. Filaments of rCF are visible as bright areas in Figure 6b, 6c and 6d. Similarly to the flax fibres, the rCF are arranged as single fibres as well as accumulations from the original tows. The fibre orientation can roughly be calculated using the minimum and maximum diameter of the ellipse. For example, in Figure 6d the group of carbon fibre marked as rCF1 has an angle of approximately 0° and is therefore oriented in the machine direction (MD = 0° laminate direction). The rCF2 in Figure 6d has a major axis length of about 40 μm and a minor axis length of about 7 μm (i.e., the diameter of the carbon fibre). Using the cosine function this results in an approximately 75° fibre angle. However, the curvature (Figure 2) of the fibres makes clear statements on fibre orientation difficult.

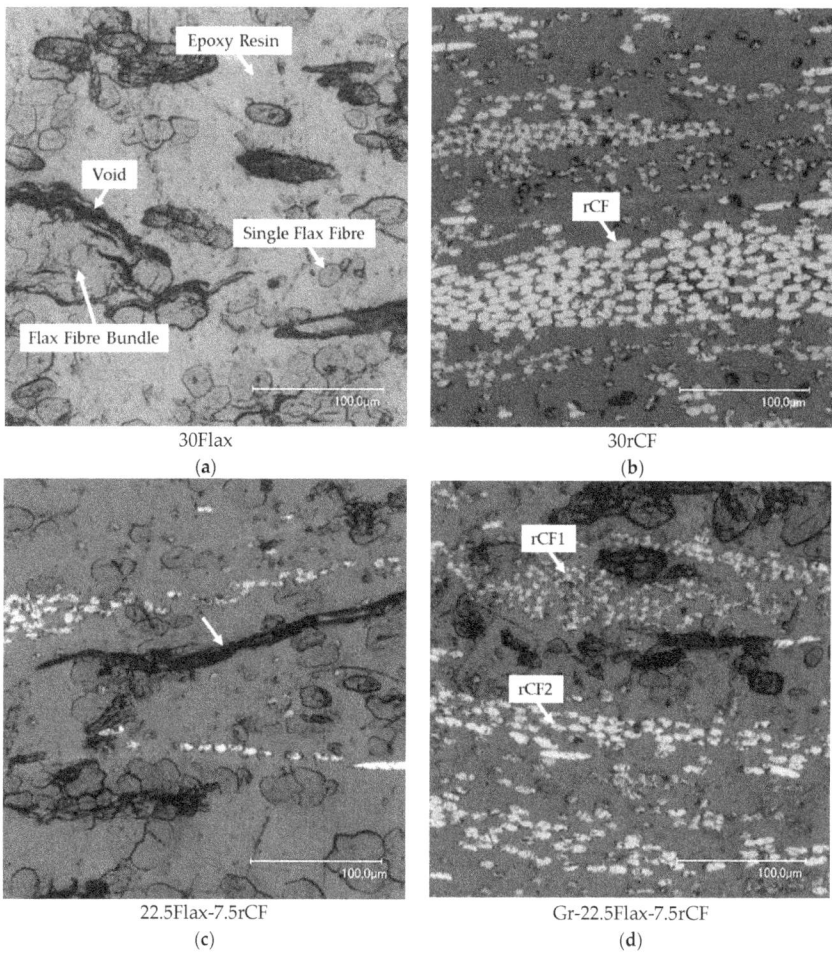

Figure 6. The micrographs show exemplary images of the fibres and their distribution in a 0° laminate direction, corresponding to the nonwoven machine direction. Reference composites of flax and rCF are shown in (**a**) respectively (**b**). All images show areas near the upper or lower surface of the composites in order to give a better distinction of the amount of rCF relative to flax fibres in the homogeneous hybrid composite (**c**) compared to the concentration of rCF in the outer layers (**d**). Voids between fibre and matrix are visible as dark areas. The elongated dark areas, as indicated by the arrow in (**c**) are probably flax fibres oriented near 90° laminate direction that were removed during the polishing process of the microscopy specimens.

The micrographs in Figure 6 indicate the relatively low fibre volume content of approximately 30% for all tested laminates, visible by major areas of epoxy resin. The calculated void content was highest for the pure flax fibre reinforced sample 30Flax (7.7%, see Table 2), while the two hybrid laminates had an even lower void content compared to the pure rCF reinforced laminate 30rCF. For the laminates containing flax fibre, the void content can partly be explained by water in the natural fibres. Generally, also the processing parameters during the laminate manufacturing have an impact on the void content [36].

2.5. Characterisation Methods

Micrographs were prepared with VFX1000 digital microscope (Keyence, Osaka, Japan) and VH-Z100UR lens (Keyence, Osaka, Japan). For this, samples in the 0° and 90° direction measuring 15 mm width were cut and embedded in a room temperature curing acrylic resin ClaroCit (Struers GmbH, Willich, Germany) followed by a three-stage polishing process. ImageJ software (version 1.50e) was used to analyse the micrograph images.

Three-point flexural tests were carried out according to the standard DIN EN ISO 14125 (class II, l/h ≅ 20) on a Zwick Roell universal testing machine Z005. A strain rate of 2 mm/min was used with a load cell of 5 kN (type KAF-Z, Zwick Roell GmbH & Co KG, Ulm, Germany). Strain was measured by cross-head displacement. Test samples were sawed to 60 mm length and 15 mm width in the 0° and 90° laminate direction on a KKS 1300 C (MAIKO Engineering GmbH). A span-to-depth ratio of 20:1 at a laminate thickness of 3 mm was chosen in this study because of the limited available number of specimens. The relatively low span-to-depth ratio can cause shear stresses resulting in additional displacements, and therefore, a potentially lower modulus [12]. Six samples were tested in the 0° direction. Because of limited composite space, a minimum of just two samples were tested in the 90° direction. The flexural modulus was calculated with the secant method between 0.05% and 0.25% strain.

3. Results and Discussion

The mean results and standard deviations on flexural strength, modulus and strain at failure obtained from the flexural test are summarized in Table 3. Figure 7 provides a graphical overview of the test results for flexural strength and modulus. It can be observed that the fibre type and distribution in hybrid laminates has a strong impact on the flexural mechanical properties.

Table 3. Data for flexural strength (σ), flexural modulus (E) and strain at break (ε) testing in 0° and 90° laminate directions with standard deviations. All layers of the laminates are pointing in the same direction. 0° corresponds to the machine direction (MD) in nonwoven. 90° corresponds to the cross direction (CD) in nonwoven.

Laminate	Flexural Strength [MPa]		Flexural Modulus [MPa]		Strain at Failure [%]	
	0°	90°	0°	90°	0°	90°
30Flax	131.2 ± 7.0	85.6 ± 3.2	7902 ± 385	4363 ± 141	2.92 ± 0.08	2.93 ± 0.06
30rCF	491.5 ± 44.6	279.9 ± 39.9	22795 ± 2367	11736 ± 1113	2.00 ± 0.17	2.35 ± 0.16
22.5Flax-7.5rCF	225.3 ± 15.0	114.2 ± 9.4	12540 ± 643	5562 ± 652	2.25 ± 0.18	2.54 ± 0.09
Gr-22.5Flax-7.5rCF	286.7 ± 18.8	165.5 ± 14.2	16818 ± 1987	7763 ± 903	1.82 ± 0.16	2.37 ± 0.07

As a reference, the pure flax fibre reinforced laminate 30Flax achieved the lowest flexural mechanical properties of all tested laminates. Mean flexural strength was 131.2 MPa in the 0° laminate direction, while the mean flexural stiffness was 7.9 GPa. A decrease of 35% for flexural strength and 45% for flexural modulus was found in the transverse (90°) direction compared to the 0° laminate direction. The lower flexural properties in 90° can be explained by the stronger alignment of the fibres in the machine direction (MD) compared to the cross direction (CD) for the nonwoven produced with the electrical carding machine (see Section 2.2).

The laminate with the highest flexural modulus and strength was reinforced purely with rCF nonwoven (30rCF). In the 0° laminate direction, 30rCF reached a mean flexural strength of 491.5 MPa and a modulus of 22.8 GPa. Flexural strength and modulus dropped by 43% and 49%, respectively, in the 90° laminate direction. Compared to the pure flax fibre reinforced variant (30Flax), a strong increase of 274% for flexural strength and 188% for flexural modulus was observed for 30rCF.

The intralayer hybrid laminate 22.5F-7.5rC, with an rCF to flax ratio of 3:1 distributed evenly in the laminate, reached a flexural strength of 225.3 MPa and a flexural modulus of 12.5 GPa in the 0° direction. Similar to the other laminates, flexural strength and modulus decreased by 49% and 56%, respectively, in the 90° laminate test direction. Compared to the pure flax reinforced laminate 30F, the substitution of 7.5 vol.-% flax fibres by rCF led to an increase of 72% strength and 59% stiffness in the 0° direction. The increase in flexural modulus and flexural strength was considerably lower in the 90° laminate test direction compared to the 0° direction. This corresponds to the stronger alignment of fibres in the machine direction (MD) during manufacturing of the nonwoven and led to a stronger influence of the matrix in the 90° direction.

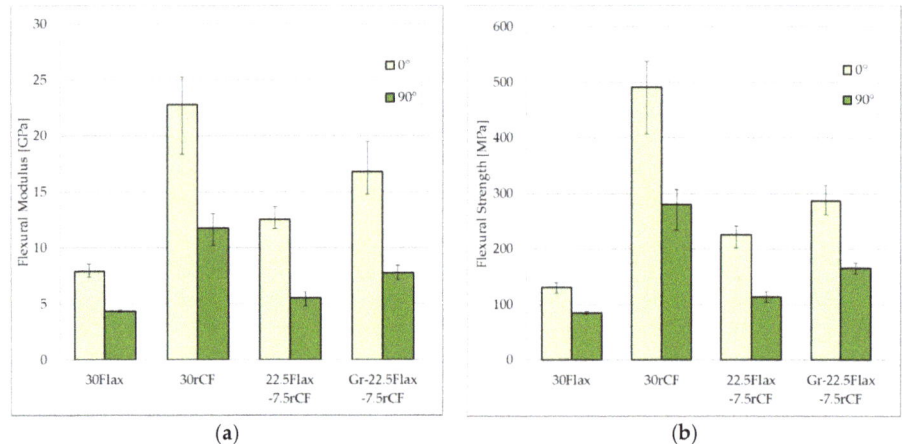

Figure 7. Flexural modulus (**a**) and flexural strength (**b**) in the 0° and 90° composite direction in accordance with DIN EN ISO 14125. The error bars show the minimum and maximum values measured for each set of specimens.

In contrast to 22.5Flax-7.5rCF, the second hybrid laminate Gr-22.5Flax-7.5rCF contained the same total volumetric flax to rCF ratio of 3:1. However, here all rCF fibres were concentrated in the outer layers with flax to rCF ratio of 1:3. All remaining flax fibres were located as non-hybrid layers in the four middle layers (see Table 1). This resulted in a combined intralayer and interlayer hybrid configuration. Concentrating the rCF in the outer layers resulted in a flexural strength of 286.7 MPa and a flexural modulus of 16.8 GPa in the 0° laminate direction. Compared to the evenly distributed rCF/flax mix in the laminate 22.5Flax-7.5rCF, the flexural strength and modulus were further increased by 27% and 34%, respectively. Compared to the pure flax fibre reinforced variant (30Flax), an increase of 118% flexural strength and 113% modulus was found. The flexural properties of the pure rCF laminate (30rCF) were still out of reach with a reduction of flexural strength of −71% for Gr-22.5Flax-7.5rCF. However, the flexural modulus of the Gr-22.5Flax-7.5rCF laminate approaches the stiffness of the 30rCF laminate with a loss of −36%. A reduction in flexural strength (−42%) and modulus (−54%) in the 90° laminate direction compared to 0° was found for Gr-22.5Flax-7.5rCF.

Typical stress-strain curves for each set of specimens in the 0° laminate direction (=machine direction of nonwoven) can be seen in Figure 8, while Figure 9 gives an overview of the fracture pattern after completion of the three-point bending tests. All measurement curves show an almost linear, slightly degressive behavior. Major differences in the stress-strain curves can be observed in stiffness, stress and strain at failure. While the 30rCF laminate obtained the highest flexural stress at failure, it also had the steepest rise, resulting in the highest stiffness. The fracture pattern of 30rCF is shown in Figure 9b. Fracture occurred abruptly on the tension side (lower side in the specimen), with clearly visible delamination effects and branched crack patterns. The reference laminate 30Flax, made purely of flax fibres, had the lowest mechanical properties of approximately 125 MPa flexural strength and 8 GPa flexural modulus. A more degressive behaviour compared to 30rCF is visible for the test curve in Figure 8. The strain at failure of about 3% is considerably higher compared to the laminates containing rCF with a strain at failure of up to 2.5%. Just as in laminate 30rCF, the fracture of 30Flax occurred on the tension side of the laminate, without observation of compression damage (Figure 9a).

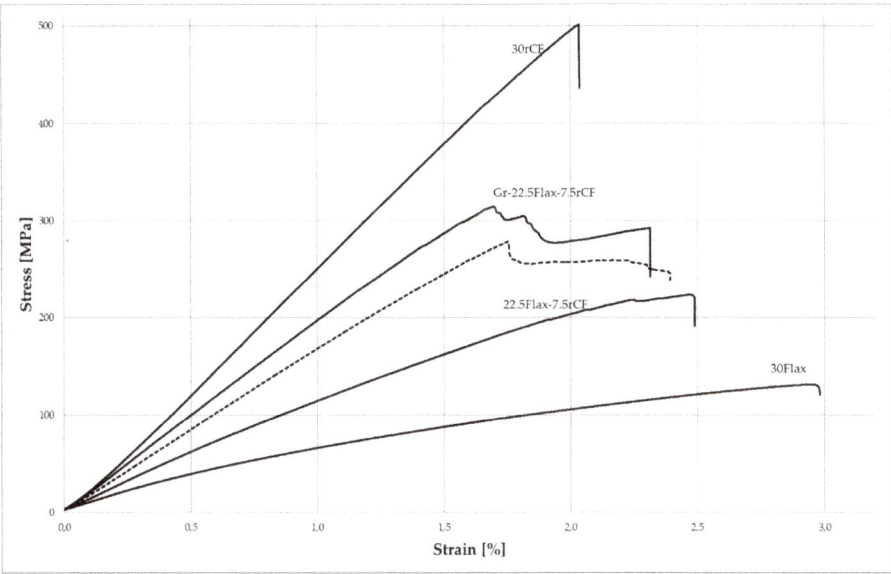

Figure 8. Stress-strain curves of three-point bending test samples in the 0° laminate test direction (similar to the machine direction of the nonwoven). For the sake of clarity, only one typical curve, with results comparable to the average values per set of specimens, is shown here. For Gr-22.5Flax-7.5rCF, a second dashed curve is indicating the considerable deviations between the test specimens because of the manual nonwoven production.

The hybridisation of flax and rCF introduced further variables, which resulted in a more complex behaviour of the stress-strain curves in the form of several drops of load (primary and final failure). The hybrid laminate 22.5Flax-7.5rCF (with evenly distributed rCF) shows several minor drops, starting at a strain of approximately 1.9%. Each drop is followed by a further increasing stress level as can be seen in the exemplary measurement curve in Figure 8. Similar to the purely flax and rCF reinforced laminates, the fracture pattern was predominantly visible on the tension side (Figure 9c).

A stronger distinction between primary and final flexural failure was observed for the gradient laminate Gr-22.5Flax-7.5rCF. Kretzis et al. [12] described the flexural behaviour of hybrid fibre reinforced composites. They found that for some hybrid configurations, the stress at primary failure was about half of the stress at final failure. In the case of the laminate Gr-22.5Flax-7.5rCF, the first failure occurred on the compression side. The location of the compression failure (buckling) is indicated with an arrow in Figure 9d and can also be attributed to a slight drop visible in the exemplary stress-strain curve at a strain of approximately 1.4% (Figure 8). This first compressive failure had no significant effect on the further development of the stress-strain curve. Final failure started on the tension side, comparable to the other tested laminates. However, the final failure did not occur as a single catastrophic failure. Instead, after the first visible tensile failure, the stress level dropped slightly at a strain of approximately 1.7%, until the specimen failed completely at a strain of approximately 2.3%.

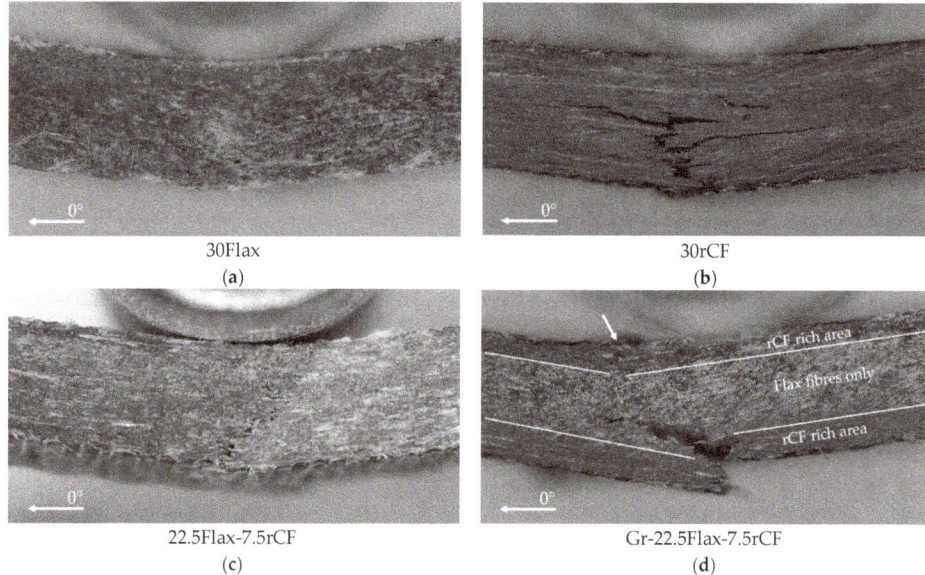

Figure 9. Fracture patterns after completion of the three-point bending tests in the 0° laminate test direction. The arrow in (**d**) marks the first area of compressive failure in the laminate with rCF concentration in the outer layers while the final failure occurred on the tensile side. The rCF rich areas in the laminate Gr-22.5Flax-7.5rCF (**d**) are marked. All other samples (**a**–**c**) failed on the tensile side without any visible compression damage. All images were extracted from video footage taken during testing with Panasonic GH-5 and Sigma 105 mm/2.8 DG Macro HSM lens mounted on a tripod. The loading pin of the flexural test is the blurred object in the upper part of the images.

In the 90° laminate test direction the flexural mechanical properties were considerably lower than the 0° test direction because of the main fibre alignment in the machine direction of the composite. The failure in the 90° laminate test direction always occurred on the tension side without visible compression failure. Typically, the crack pattern was less branched compared to the 0° laminate test direction, with the exception of the 30rCF laminate, which showed a similar crack pattern in the 0° and 90° test direction.

Deviations in flexural mechanical properties measured for the hybrid nonwoven specimens were observed. The highest standard deviation in the 0° laminate test direction was found in the Gr-22.5Flax-7.5rCF variant. In principle, the standard deviation varies between 4.5% and 11.8% in the 0° laminate test direction. Two curves measured during the flexural test for the Gr-22.5Flax-rCF samples are shown in Figure 8 as an example. The dashed line shows a low performing sample of Gr-22.5Flax-7.5rCF for comparison. Different reasons for the standard deviation can be considered and will be discussed in the following paragraph.

The distribution of fibres (rCF and flax) in the nonwoven was not fully homogeneous and led to variance in the fibre volume content in the laminate. For example, relative accumulations of rCF lead to higher mechanical properties in these specific areas compared to areas with low fibre content. An example can be observed in Figure 5c, which shows a deviation in the thickness of the rCF rich outer layers. The random distribution of fibres in the nonwoven also has an effect on the fibre orientation in different areas of the laminate. This can lead to fibres oriented anywhere between the 0° and 90° laminate direction. The laboratory nonwoven produced in this study also showed a strong curvature of fibres which resulted in reduced mechanical performance compared to perfectly aligned fibres. Another aspect, generally applicable for natural fibre reinforced polymers, is the embedding of shives into the composite and a considerable void content. Shives are wooden particles that can be left over from the manufacturing process of the flax fibres. With a possible shive diameter in the millimetre range, compared to approximately 20 μm for flax fibres, these shives can act as a weak point in the composite. Voids and unwanted constituents like shives can increase stresses locally, leading to earlier failure [37].

4. Conclusions and Outlook

Growing environmental concerns have led to an increased interest in the application of bio-based materials such as flax fibres in composites. In parallel, increasing amounts of CFRP waste raise questions on the utilisation of valuable recycled carbon fibres (rCF). Recycling only makes sense when the regained materials are used again in new applications. As NFRP usually have low mechanical properties compared to GFRP and CFRP, a possible solution for the improvement could be the addition of rCF in a hybrid composite. This study gives a preliminary insight into the potential of mixing flax fibres with rCF in a hybrid nonwoven and the effect on the flexural properties of epoxy composites.

The nonwoven, containing flax and rCF was manufactured manually with a two-drum card. While obtaining the typical overall randomness of fibre distribution, the nonwoven had a fibre orientation in the machine direction. Epoxy laminates with 30% fibre volume content were manufactured. The results presented in this study show that mixing natural bast fibres such as flax with a small amount of recycled carbon fibres (7.5 volume percent of the laminate) can lead to a significant increase in flexural mechanical properties compared to a composite only reinforced with flax fibres. The concentration of rCF in the outer layers of the laminate, away from the neutral axis, further increased the flexural mechanical properties, compared to the same mixing ratio of rCF to flax distributed evenly in the laminate. However, the flexural mechanical properties of composites purely reinforced with rCF were not obtained with the hybrid variants.

The hybrid combination of flax and rCF makes it possible to modify the material characteristics. The mixing ratio and distribution of the different fibres can be fully adapted to the requirements of the desired application. While such hybrid composites lose their fully bio-based nature, they may gain enough multifunctional characteristics to compete with fully synthetic materials. Usually, a high degree of freedom for the choice of fibre orientation, ranging from a strong fibre orientation up to a more isotropic material behaviour, can be obtained by standard nonwoven manufacturing processes. Therefore, a hybrid combination of natural fibres and recycled carbon fibres can be considered as a possible alternative to state-of-the-art materials such as GFRP for application in semi-structural applications. However, a more detailed characterisation of the hybrid nonwoven from rCF and flax is required. Tensile, compressive, damping and fatigue behaviour need to be characterised for a better understanding of the potential of natural fibres combined with recycled carbon fibres. An assessment of the damping properties is important to highlight the multifunctional advantages of the multiscale structure of natural fibres. Furthermore, recycled carbon fibres from a pyrolysis process should also be considered. Such rCF are decreased in size by the pyrolysis step, usually leading to reduced fibre-matrix adhesion. Chemical and physical treatments of the rCF could help to improve this interface.

Finally, it must be stressed that the environmental impacts need to be assessed. For example, a reduction in environmental impacts is not generally possible by just substituting synthetic fibres with natural fibres. The use-phase in transportation applications usually has a strong impact on the environmental properties. Another important aspect is the end-of-life treatment. NFRP can be incinerated but the addition of rCF would make another process necessary. Further recycling of the rCF in the hybrid laminates is theoretically possible. Such "re-recycled" rCF could be recovered in an even more deteriorated quality and used to produce milled rCF in the form that is already available today [38]. It is important to calculate the environmental impacts of all life phases of composites with a life cycle assessment (LCA); from raw material extraction, production, and the use-phase to the end-of-life treatments, and to compare them with competing materials.

Author Contributions: J.B. conceived and designed the experiments; J.B. performed the experiments; J.B., P.W. and M.W. analysed the data; J.B. wrote the paper.

Funding: This project has received funding from the European Union's Horizon 2020 research and innovation programme under grant agreement No 690638 and the Special Research Plan on Civil Aircraft of Ministry for Industry and Information of the People's Republic of China (MIIT) under Grant No MJ-2015-H-G-103.

Conflicts of Interest: The authors declare no conflict of interest.

References

1. Bachmann, J.; Yi, X.; Gong, H.; Martinez, X.; Tserpes, K.; Ramon, E.; Paris, C.; Moreira, P.; Fang, Z.; Li, Y.; et al. Outlook on ecologically improved composites for aviation interior and secondary structures. *CEAS Aeronaut. J.* **2018**, *9*, 533–543. [CrossRef]

2. Li, Y.; Hi, Y.; Hu, C.; Yu, Y. Microstructures and mechanical properties of natural fibers. *Adv. Mater. Res.* **2008**, *33–37*, 553–558. [CrossRef]

3. Pickering, K.L.; Aruan Efendy, M.G.; Le, T.M. A review of recent developments in natural fibre composites and their mechanical performance. *Compos. A* **2016**, *83*, 98–112. [CrossRef]

4. Yan, L.; Chouw, N.; Jayaraman, K. Flax fibres and its composites—A review. *Compos. B* **2014**, *56*, 296–317. [CrossRef]

5. Summerscales, J.; Dissanayake, N.; Virk, A.; Hall, W. A review of bast fibres and their composites. Part 2—Composites. *Compos. B* **2010**, *41*, 1336–1344. [CrossRef]

6. Wambua, P.; Ivens, J.; Verpoest, I. Natural fibres: Can they replace glass in fibre reinforced plastics? *Compos. Sci. Tech.* **2003**, *63*, 1259–1264. [CrossRef]

7. Satyanarayana, K.G.; Arizaga, G.G.C.; Wypch, F. Biodegradable composites based on lignocellulosic fibers —An overview. *J. Prog. Polym. Sci.* **2009**, *34*, 982–1021. [CrossRef]

8. Sood, M.; Dwivedi, G. Effect of fiber treatment on flexural properties of natural fiber reinforced composites: A. *review. Egypt. J. Pet.* **2017**. [CrossRef]

9. Xie, Y.; Hill, C.A.S.; Xiao, Z.; Militz, H.; Mai, C. Silane coupling agents used for natural fiber/polymer composites: A review. *Compos. A* **2010**, 806–819. [CrossRef]

10. Li, Y.; Yi, X.; Xian, G. An overview of structural-functional-integrated composites based on the hierarchical microstructures of plant fibers. *Adv. Compos. Hybrid. Mater.* **2018**, *1*, 232–246. [CrossRef]

11. Swolfs, Y.; Crauwels, L.; Van Breda, E. Tensile behaviour of intralayer hybrid composites of carbon fibre and self-reinforced polypropylene. *Compos. A* **2014**, *59*, 78–84. [CrossRef]

12. Kretzis, G. A review of the tensile, compressive, flexural and shear properties of hybrid fibre-reinforced plastics. *Comoposites* **1987**, *18*, 13–23. [CrossRef]

13. Park, R.; Jang, J. Effect of laminate geometry on impact performance of aramid fiber/polyethylene fiber hybrid composites. *J. Appl. Polym. Sci.* **2000**, *75*, 952–959. [CrossRef]

14. Pegoretti, A.; Fabbri, E.; Migliaresi, P.F. Intraply and interplay hybrid composites based on E-glass and poly(vinyl alcohol) woven fabrics: Tensile and impact properties. *Polym. Int.* **2004**, *53*, 1920–1927. [CrossRef]

15. Manders, P.W.; Bader, M.G. The strength of hybrid glass/carbon fibre composites. *J. Mater. Sci.* **1981**, *16*, 2233–2245. [CrossRef]

16. Indra Reddy, M.; Anil Kumar, M.; Rama Bhadri Raju, C. Tensile and Flexural properties of Jute, Pineapple leaf and Glass Fiber Reinforced Polymer Matrix Hybrid Composites. *Mater. Today* **2018**, *5*, 458–462. [CrossRef]

17. Lützkendorf, R.; Reussmann, T.; Danzer, M. Hybride Verbundwerkstoffe mit recycelten Carbonfasern. *Lightweightdesign* **2017**, *2*, 22–26. [CrossRef]

18. Flynn, J.; Amiri, A.; Ulven, C. Hybridized carbon and flax fiber composites for tailored performance. *J. Mat. Des.* **2016**, *102*, 21–29. [CrossRef]

19. Adekunle, K.; Cho, S.W.; Ketzscher, R.; Skrifvars, M. Mechanical Properties of Natural Fiber Hybrid Composites Based on Renewable Thermoset Resins Derived from Soybean Oil, for use in technical applications. *J. Appl. Polym. Sci.* **2012**, *124*, 4530–4541. [CrossRef]

20. Cicala, G.; Pergolizzi, E.; Piscopo, F.; Carbone, D.; Recca, G. Hybrid composites manufactured by resin infusion with a fully recyclable bioepoxy resin. *Compos. B* **2018**, *132*, 69–79. [CrossRef]

21. Murdani, A.; Hadi, A.; Amrullah, U.S. Flexural Properties and Vibration Behavior of JuteGlassCarbon Fiber Reinforced Unsaturated Polyester Hybrid Composites for Wind Turbine Blade. *Key Eng. Mater.* **2017**, *748*, 62–68. [CrossRef]

22. Witten, E.; Kraus, T.; Kühnel, M. Composites-Marktbericht 2015: Marktentwicklungen, Trends, Ausblicke und Herausforderungen. CCeV and AVK, 21 September 2015. Available online: http://www.avk-tv.de/files/20151214_20150923_composites_marktbericht_gesamt.pdf (accessed on 25 June 2018).

23. Yang, Y.; Boom, R.; Irion, B.; van Heerden, D.J.; Kuiper, P. Recycling of composite materials. *Chem. Eng. Process.* **2012**, *51*, 53–68. [CrossRef]

24. Carberry, W. Airplane Recycling Efforts Benefit Boeing Operators. Available online: http://www.boeingvideo.com/commercial/aeromagazine/articles/qtr_4_08/pdfs/AERO_Q408_article02.pdf (accessed on 30 June 2018).

25. Verma, S.; Balasubramamaniam, B.; Gupta, R.K. Recycling, reclamation and re-manufacturing of carbon fibres. *Curr. Option Green Sus. Chem.* **2018**, *13*, 86–90. [CrossRef]

26. Oliveux, G.; Dandy, L.O.; Leeke, G.A. Current status of recycling of fibre reinforced polymers: Review of technologies, reuse and resulting properties. *Prog. Mater. Sci.* **2015**, *72*, 61–99. [CrossRef]

27. Naqvi, S.R.; Mysore Prabhakara, H.; Bramer, E.A.; Dierkes, W. A critical review on recycling of end-of-life carbon fibre/glass fibre reinforced composites waste using pyrolysis towards a circular economy. *Resour. Conserv. Recycl.* **2018**, *136*, 118–129. [CrossRef]

28. Rybicka, J.; Tiwari, A.; Leeke, G.A. Technology readiness level assessment of composites recycling technologies. *J. Clean. Prod.* **2016**, *112*, 1001–1012. [CrossRef]

29. Gardiner, G. Recycled Carbon Fiber Update: Closing the CFRP Lifecycle Loop. *Compos. Tech.* **2014**. Available online: https://www.compositesworld.com/articles/recycled-carbon-fiber-update-closing-the-cfrp-lifecycle-loop (accessed on 25 June 2018).

30. Jesson, D.A.; Watts, J.F. The Interface and Interphase in Polymer Matrix Composites: Effect on Mechanical Properties and Methods for Identfication. *Polym. Rev.* **2012**, *52*, 321–354. [CrossRef]

31. Fischer, H.; Schmid, H.G. Quality Control for Recycled Carbon Fibres. *Kunststoffe Intern.* **2013**, *11*, 68–71.

32. Illing-Günther, H.; Hofmann, M.; Gulich, B. Nonwovens made of recycled carbon fibres as basic material for composites. In Proceedings of the 7th International CFK-Valley Stade Convention "Latest Innovations ins CFPR Technology", Stade, Germany, 11–12 June 2013.

33. Recycling Today. ELG Carbon Fibre Ltd. to Highlight Role of Recycled Carbon Fibre. 2016. Available online: http://www.recyclingtoday.com/article/elg-carbon-fibre-jec-world-2016-exhibit (accessed on 31 August 2018).

34. Albrecht, W.; Fuchs, H.; Kittelmann, W. *Nonwoven Fabrics: Raw Materials, Manufacture, Applications, Characteristics, Testing Processes*; Wiley-VCH Verlag GmbH & Co. KGkA: Germany, 2015; Available online: https://onlinelibrary.wiley.com/doi/book/10.1002/3527603344 (accessed on 30 June 2018).

35. Lütke, C.; Rübsam, U.; Schlüter, T.; Schröter, A.; Gloy, Y.S. Sustainability in Luxury Textile Applications: A Contradiction or a New Business Opportunity. In *Handbook of Sustainable Luxury Textiles and Fashion*; Gardetti, M.A., Muthu, S.S., Eds.; Springer: Singapore, 2015; pp. 121–143. ISBN 978-981-287-632-4.

36. Li, Y.; Ma, H.; Shen, Y.; Li, Q.; Zheng, Z. Effects of resin inside fiber lumen on the mechanical properties of sisal fiber reinforced composites. *Compos. Sci. Technol.* **2015**, *108*, 32–40. [CrossRef]

37. Pupure, L.; Varna, J.; Joffe, R.; Berthold, F.; Miettinen, A. Mechanical properties of natural fiber composites produced using dynamic sheet former. *Wood Mater Sci. Eng.* **2018**. [CrossRef]

38. Composites World 2016. Recycled Carbon Fiber: Its Time Has Come. Available online: https://www.compositesworld.com/columns/recycled-carbon-fiber-its-time-has-come (accessed on 8 September 2018).

Article

A Multi-Scale Modeling Approach for Simulating Crack Sensing in Polymer Fibrous Composites Using Electrically Conductive Carbon Nanotube Networks. Part II: Meso- and Macro-Scale Analyses

Konstantinos Tserpes * and Christos Kora

Laboratory of Technology & Strength of Materials, Department of Mechanical Engineering & Aeronautics, University of Patras, Patras 26500, Greece; chriskora93@gmail.com
* Correspondence: kitserpes@upatras.gr

Received: 26 August 2018; Accepted: 4 October 2018; Published: 9 October 2018

Abstract: This is the second of a two-paper series describing a multi-scale modeling approach developed to simulate crack sensing in polymer fibrous composites by exploiting interruption of electrically conductive carbon nanotube (CNT) networks. The approach is based on the finite element (FE) method. Numerical models at three different scales, namely the micro-scale, the meso-scale and the macro-scale, have been developed using the ANSYS APDL environment. In the present paper, the meso- and macro-scale analyses are described. In the meso-scale, a two-dimensional model of the CNT/polymer matrix reinforced by carbon fibers is used to develop a crack sensing methodology from a parametric study which relates the crack position and length with the reduction of current flow. In the meso-model, the effective electrical conductivity of the CNT/polymer computed from the micro-scale is used as input. In the macro-scale, the final implementation of the crack sensing methodology is performed on a CNT/polymer/carbon fiber composite volume using as input the electrical response of the cracked CNT/polymer derived at the micro-scale and the crack sensing methodology. Analyses have been performed for cracks of two different lengths. In both cases, the numerical model predicts with good accuracy both the length and position of the crack. These results highlight the prospect of conductive CNT networks to be used as a localized structural health monitoring technique.

Keywords: carbon nanotubes; polymer nanocomposites; electrical conductivity; crack sensing; multi-scale modeling

---

1. Introduction

Structural Health Monitoring (SHM) aims to provide at every moment during the life of a structure, a diagnosis of the "integrity" of the materials, of the different structural parts, and of the entire structure. Based on Reference [1], an effective SHM system aims to minimize the time required for ground inspections, to increase the operation time of the aircraft and to reduce the maintenance cost by more than 30%. These advantages represent a major contribution towards a greener aviation. SHM has two modules, i.e., the diagnosis and prognosis. The diagnosis module is based on a monitoring system, which usually consists of a network of sensors distributed over a relatively large area of the structure. However, with such a monitoring system only major damage events can be detected. The current trend is the development of dense wireless networks of very small sensors. Moreover, multifunctional materials for SHM have gained attention due to their ability to sense, actuate and harvest energy from ambient vibrations.

Carbon nanotubes (CNTs) are used as nanofillers to produce multifunctional polymers (nanocomposites) and fiber composites due to their extraordinary mechanical [2–7], thermal [8]

and electrical properties [2,7,9] combined to their 1D structure. Amongst the targeted functionalities of the CNT-based materials is strain and damage sensing. Damage sensing in fiber composites by conductive CNT networks has been initially studied by Fiedler et al. [10]. Since then, numerous works, mainly experimental (e.g., References [11–15]), have been performed. On the contrary, there have been reported only a few models of damage sensing by CNT networks. The most representative of them are those of References [13] and References [14]. In [13], the authors modeled damage sensing in [0/90]s cross-ply glass fiber composites using embedded CNT networks. The contact resistances of the CNTs were modeled using the electrical tunneling effect and the effective electrical resistance of the percolating CNT network was calculated by considering nanotube matrix resistors and employing the FE method for electrical circuits. The loading process of the composite, from initial loading to final failure, was simulated also by the FE method. The deformation and damage induced resistance change was identified at each loading step. The numerical results show that the model captures the essential parameters influencing the electrical resistance of the CNT networks. The authors did not carry out any parametric study. In [14], the authors presented an analytical model of the strain sensing behavior of CNT-based nanocomposites. The model incorporated the electrical tunneling effect due to the matrix material between the CNTs to describe the electrical resistance variation due to mechanical deformation. The model simulates the inter-nanotube matrix deformation at the micro/nanoscale due to the macroscale deformation of the nanocomposites. A comparison of the analytical predictions with the experimental data showed that the proposed model simulates the sensing behavior efficiently.

Based on the above overview, we can summarize that the understanding of the key factors governing strain and damage identification mechanisms of nanocomposites through localized variations of the electrical conductivity in the absence of mechanical loads has been mainly attempted experimentally. However, in the design of such SHM systems, simulation-driven design tools could be very useful as they could to the reduction of development cost and time. Within this framework, in the present paper, proposed is a multi-scale FE-based modeling approach for simulating the basic mechanisms of crack sensing by conductive CNT networks in nanocomposites and fibrous composites. The description of the modeling approach is done in two papers. The first paper [15] describes the micro-scale model while the present paper (second paper) describes the development of a crack sensing methodology by means of a meso-scale model, a parametric study and the final implementation of the crack sensing methodology.

2. Meso-Scale Model

2.1. Geometry and FE Model

In the meso-scale, a 2D model of a CNT/polymer reinforced by carbon fibers has been developed. The modeled area is a transverse section of size of 0.2 mm × 0.2 mm containing 400 carbon fibers (circular cross-sections). The fibers are placed in a square configuration, but they are not equally spaced and do not have the same diameter. The model was structured in line with electron microscopy images of unidirectional carbon fiber reinforced plastic (UD CFRP) cross sections which show a random positioning and diameter of carbon fibers. In the process of developing a realistic approach, this observation is incorporated as a geometrical variation. The model describes a quadratic distribution and is divided in square fiber-polymer sections of size of 10 μm × 10 μm. In each sector, the fibers' diameter varies according to $D = rand(6,8)$ (μm), thus a unique value inside this interval is assigned iteratively. Simultaneously, there is shift of the fiber center from the center of the section in the x and y axes given by $x = rand(1.3 * (D/2), s - 1/3(D/2))$ and $y = rand(1.3 * (D/2), s - 1/3(D/2))$ where s is the section's size and D is the diameter value at the given iteration. The variation quantities are arbitrary with respect to the physical mechanisms though they were chosen based on the requirement of the minimum mesh distortion.

The aim of this analysis is to develop a crack sensing methodology through the parametric correlation of current drop with the characteristics of the crack (length and position). The analysis is based on the simplicity of the implementation of the direct current (DC) for measuring the variation of resistance in conductive composites (e.g., the four-probe method e.g., Reference [11]) and the through-the-thickness measurement of resistance in thin-walled structures with a known transverse conductivity. Both the fibers and the matrix are modeled using the 2D 8-node ANSYS PLANE223 triangular elements [16]. The 3-node option of the elements has been used as it was found that it reduces significantly the computational effort without compromising considerably the accuracy of the results. PLANE223 has many capabilities including structural-thermal and thermal-electric capabilities. The element input data includes eight nodes and structural, thermal, and electrical material properties. The FE mesh of the meso-scale model is shown in Figure 1.

For the transition to the meso-scale, we consider electrically homogeneous matrix and a full bonding between the fibers and the matrix. The transition from micro-scale to meso-scale is achieved through the definition of discrete electrical conductivity regions using the values computed in the validation process of the micro-scale model as demonstrated in the first part [15] for the 5 volume fractions above the percolation threshold. The regions were distributed randomly utilizing APDL algorithm to alter element properties and this was done to reproduce the inhomogeneity of network density.

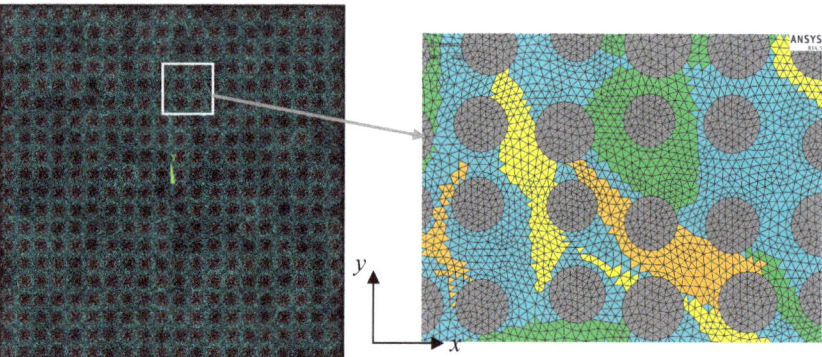

Figure 1. FE mesh of the meso-scale model. The different colors in the elements of the matrix indicate areas with different effective electrical conductivity due to the various density of the CNT network that comes from the micro-scale.

2.2. Imposed Electric Field

Charging has been applied to the model using a series of circuit ANSYS CIRCU124 elements [16] placed at the top nodes. These are 2-node circuit elements with a constant resistance which are not affected by geometrical factors. The contact with the upper end of the model is utilized using common nodes. A circuit element was placed at each node of the plane elements. A voltage V_0 is applied to the free node of the circuit element while the lower nodes of the model are grounded. A benefit from the use of circuit elements is the direct computation of reaction currents without the need for gathering and processing data from the interior of the model. It is noted that parasitic capacitance could occur given the miniscule distances between the resistors of the meso-scale model. However, the aim of modelling this arrangement in this scale is to bridge the microstructural electrical behavior with the macro-scale response from a resistor response standpoint. In a hypothetical implementation, the density of a resistor arrangement on a surface determines the resolution of measurement. The normalized quantities of *drop* and *span* can be used to roughly characterize discontinuities of larger scales through proportional considerations. Moreover, the use of resistor elements constitutes a modelling approach of a simplistic arrangement from which we can derive information externally from the holistic response of the model. Moreover, the data acquisition is rendered easier.

The sketch in Figure 2 describes the simple electric circuit of the meso-scale model. The reaction current i_R is simply derived from Ohm's law as

$$i_R = \frac{V_0}{R_0 + R_p},\tag{1}$$

where V_0 is the voltage and R_0 and R_p are the resistances. In the present application, R_0 is the resistance of the circuit element while R_p is the resistance of the model (material). Furthermore, the parallel network of the circuit elements provides information about the location of the current drop.

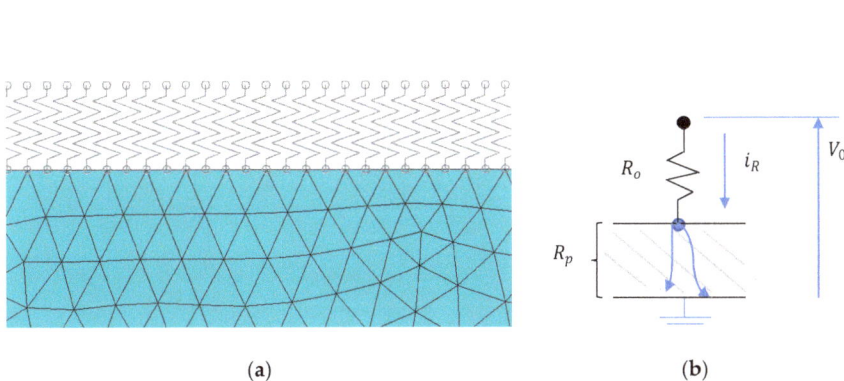

(a)   (b)

Figure 2. (a) Part of the FE mesh of the meso-scale model showing the connection of the circuit elements with the nodes of the nanocomposite and (b) sketch of the electric circuit of the meso-scale model.

2.3. Crack Sensing Methodology

The proposed crack sensing methodology is based on the parameterization of the measurements derived for different crack lengths and crack distance from the measurement points (crack depth). It is assumed that the crack is single, continuous and parallel to the x axis. Inclined cracks might be also considered by using the effective length, which is projection of the inclined crack to the x direction. The crack is introduced in the mesh by releasing nodes as it was done in the micro-scale model.

For each crack case, we derive through repetitive analyses the reaction currents $i_R^{i,l,d}$ at the circuit elements (i is the number of circuit element, l is the crack length and d is the crack depth. For a constant depth, we derive the reaction current for different crack lengths and we compare the values with the reference value of the uncracked material $i_R^{i,0}$). This way, we evaluate the effect of crack length on the current drop for each circuit through the current drop parameter $\delta i_R^{i,l,d}$ given by

$$\delta I_R^{i,l,d} = \frac{I_R^{i,0} - I_R^{i,l,d}}{I_R^{i,0}} \times 100\%,\tag{2}$$

Figure 3 shows a typical computed distribution of current drop around the crack zone. It is observed that outside the crack zone the reaction current increases (negative $\delta i_R^{i,l,d}$) due to the redistribution of current flow in the area around the crack. As shown, the maximum current drop occurs at the centroid of the crack. Using the electrical response of the reference material we define in the diagram the maximum current drop value (*drop*) and the difference between the maximum and minimum current drops (*span*). The latter was used to quantify the inhomogeneity of the electrical response, i.e., whether a sudden or a smooth increase takes place. For the difference between the maximum and minimum current drops to be generalized, *span* is normalized over the maximum current *drop* value. Based on the above, we have

$$drop = \max(\delta I_R^{i,l,d}),$$ (3)

$$span = \frac{\max(\delta I_R^{i,l,d}) - \min(\delta I_R^{i,l,d})}{\max(\delta I_R^{i,l,d})},$$ (4)

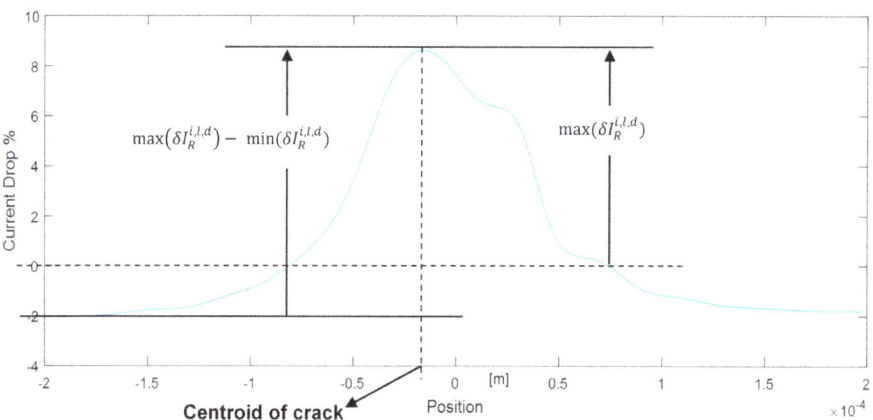

Figure 3. An example of the electrical response curve in the area around the crack.

A parametric study has been performed to correlate the different crack scenarios with the measured values of the *drop* and *span* values and develop corresponding functions. By cross-checking the *drop* and *span* values for every circuit element, the maximum values are located. Then, 2D plots of the maximum values with regards to the normalized crack length and normalized crack depth are created and every surface is fitted with a two-variable function. The outcome of this process, when compared to the outcome of the reference material can give an estimation about the size and location of the crack. Graphically, this can be implemented by using the cross-section of the curves derived from the cross-section of the *drop* and *span* surface plots as will be shown in Section 4.1.

3. Macro-Scale Model

3.1. Geometry and FE Modeling

In the macro-scale, a CNT/polymer/carbon fibers composite volume of dimensions of 10 mm × 10 mm × 2.5 mm has been modeled. The volume has been meshed with 20-node SOLID131 elements [16]. The mesh density is selected such as each element to have the dimensions of the micro-model. The FE mesh of the macro-model is shown in Figure 4. As in the mesoscale, the 5 volume fraction values used in the validation graph [15] are used as input in the full fiber-CNT-polymer microscale model and the 5 effective electrical conductivities are computed. These values are assigned randomly as the transverse electrical conductivity property of each macro-scale element using APDL code. Similarly, thee electrical conductivity of the solid elements in the fiber direction has been computed using the rule of mixtures

$$\sigma_z = \sigma_{fib} \times 0.55 + \sigma_m \times 0.45$$ (5)

in which the electrical conductivity of the matrix σ_m is either one of the 5 values in the validation graph chosen randomly by the code. In Equation (5), σ_{fib} is the conductivity of the fibers and σ_m is the conductivity of the matrix. The parameter of 0.55 corresponds to the fiber volume fraction (55%) and the 0.45 to the matrix volume fraction (%). Again, this approach aims in the numerical reproduction of the network density inhomogeneity.

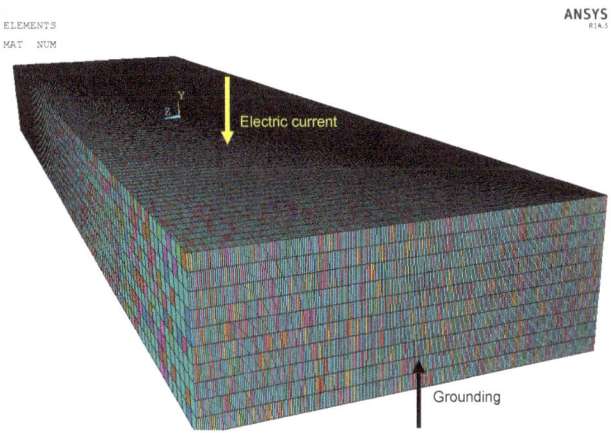

Figure 4. FE mesh of the macro-scale model.

3.2. Introduction of a Virtual Crack, Charging and Computations

As for the meso-scale model, charging has been applied to the macro-scale model using circuit elements CIRCUIT124 [16] which are placed on the nodes of the top *XY* surface while the nodes of the bottom surface are grounded. Crack growth in the composite volume has been simulated by applying the corresponding drop of effective conductivity σ_{eff} at the elements of crack path using the micro-scale analysis results on the variation of σ_{eff} with regards to crack length. Computations have been performed by gathering the values of the current from the circuit elements and comparing them with the corresponding values of the reference model. Then, the differences are qualified on the plane, thus producing a graphical representation for the location and magnitude of the current drop at the *XY* plane. In order to validate the meso-scale model, results from the macro-scale for a known crack have been used.

4. Numerical Results

4.1. Meso-Scale: Effect of Crack Presence

In the meso-scale, the electrical response of models with 39 different crack lengths and 9 different depths (from −0.1 to −0.9) has been simulated. The crack length and depth are normalized with regards to the dimension of the Representative Volume Element (RVE) [15], $l_0 = 0.2$ mm. The input parameters and material properties of the meso-scale model are listed in Table 1. Note that the resistance of the circuit elements does not affect the results.

Table 1. Input parameters and materials properties in the meso-scale model.

Parameter/Property	Value
Fiber's diameter	6–8 μm
Fiber volume fraction	55%
Fiber's electrical conductivity at Y direction, σ_{cf}^{y}	10^4 S/m
Fiber's electrical conductivity at X direction, σ_{cf}^{x}	10^4 S/m
Electrical conductivity of polymeric matrix variation, σ_m	$[3.77 \times 10^{-5}, 1.12 \times 10^{-3}, 3.28 \times 10^{-3}, 5.06 \times 10^{-3}, 8.77 \times 10^{-3}]$ S/m
Electrical resistance of circuit elements	10^3 Ohm

The predicted *drop* and *span* values for the different l_c/l_0 and d/l_0 ratios have been listed in tables. Table 2 lists the results for the case of $d/l_0 = -0.1$. The other tables are omitted for the sake of briefness. To visualize the data, 3D plots have been created from tabular data in the MATLAB software using fitting functions. Figure 5 shows the 3D plot of *drop* vs. l_c/l_0 and d/l_0 for the case of maximum *drop* while Figure 6 shows the 3D plot of span vs. l_c/l_0 and d/l_0 for the case of maximum *span*. We observe a decrease in the sensitivity of the response with increasing the depth ratio d/l_0, which means that the deeper the crack is the less sensitive the method becomes.

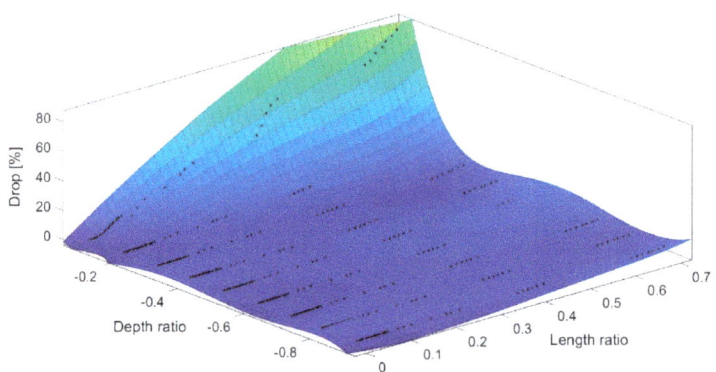

Figure 5. Fitted 3D plot of the *drop* function with regards to length ratio l_c/l_0 and depth ratio d/l_0.

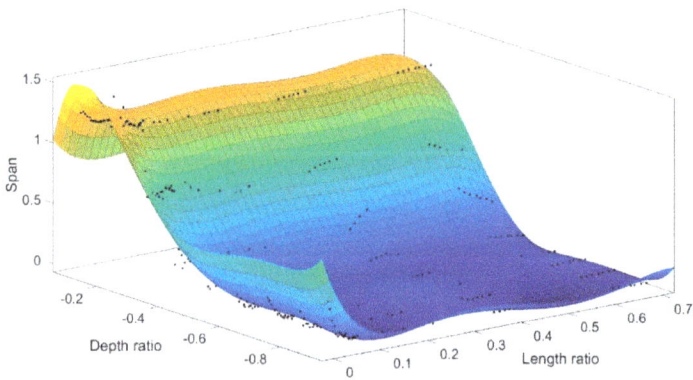

Figure 6. Fitted 3D plot of the *span* function with regards to length ratio l_c/l_0 and depth ratio d/l_0.

Table 2. *Drop* and *span* values vs. length ratio, l_c/l_0 and depth ratio $d/l_0 = -0.1$.

Length Ratio, l_c/l_0	Drop (%)	Span
0.009	0.376	1.256
0.015	0.426	1.262
0.019	0.629	1.238
0.023	0.763	1.214
0.029	1.328	1.203
0.031	1.724	1.177
0.036	2.178	1.170
0.040	3.343	1.164
0.043	3.832	1.157
0.047	4.508	1.151
0.050	5.250	1.152
0.052	6.457	1.152
0.057	7.012	1.153
0.060	7.850	1.153
0.066	8.794	1.153
0.067	9.345	1.152
0.070	9.897	1.145
0.072	10.15	1.149
0.077	10.64	1.148
0.100	8.644	1.232
0.112	12.920	1.208
0.129	17.400	1.193
0.174	21.664	1.181
0.208	27.084	1.170
0.218	29.910	1.166
0.237	32.345	1.162
0.367	35.678	1.159
0.379	38.693	1.157
0.390	44.412	1.155
0.400	51.903	1.151
0.410	56.378	1.149
0.426	59.376	1.147
0.622	65.263	1.144
0.633	67.230	1.142
0.645	71.411	1.141
0.660	74.113	1.141
0.667	77.364	1.142
0.681	79.746	1.143
0.692	83.053	1.146

The opposite procedure, i.e., the characterization of a cracked model, lies in the determination of crack length and position using the tabular data or the graphs. To demonstrate the process, we select random values of *drop* and *span*, for instance, *drop* = 6.2% and *span* = 1.18%, and from the graphs of Figures 7 and 8 using intersection lines we gather all sets of the l_c/l_0 and d/l_0 ratios which have given the selected values. The outcome of this first step are projection-curves (see Figures 7 and 8). Then, we plot the two lines in a l_c/l_0 vs. d/l_0 system (Figure 9) and we find their intersections. The intersection points give sets of l_c/l_0 and d/l_0 ratios which lead to the selected *drop* and *span* values. Since the intersections are more than one, this means we have more than one sets of *drop* and *span* values that lead to the same electrical result. The selection of a unique solution requires more data. Nevertheless if we use the solution with the maximum l_c/l_0 ratio, i.e., $l_c/l_0 = 0.28$ and $d/l_0 = -0.21$, we get a crack length of $l_c = 0.056$ mm and a depth of $d = 0.042$ mm.

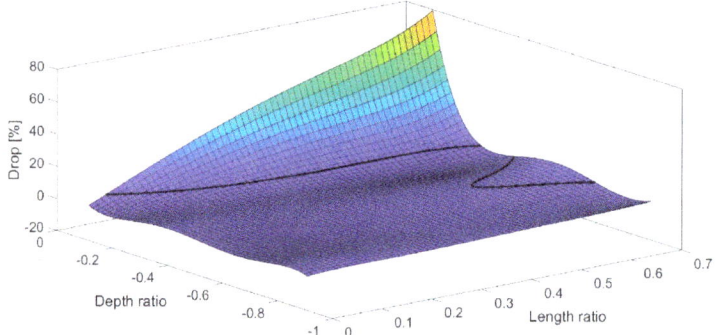

Figure 7. Intersection curve for *drop* = 6.2%.

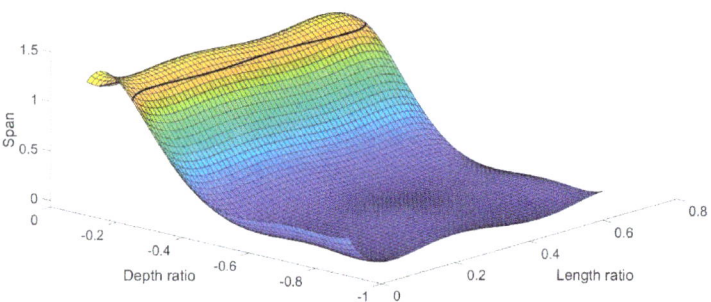

Figure 8. Intersection curve for *span* = 1.18%.

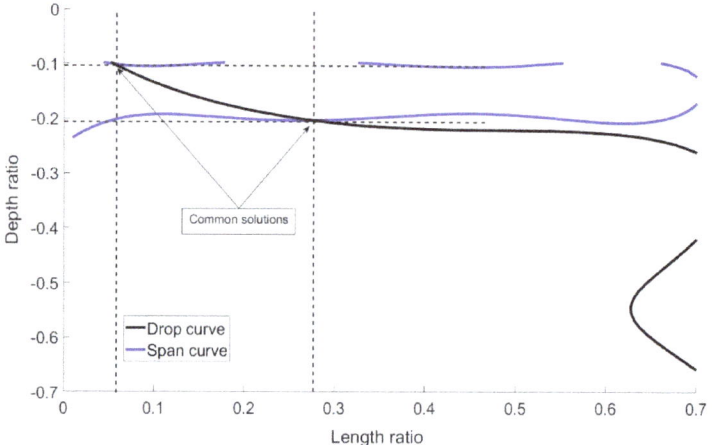

Figure 9. Intersection curves for *drop* = 6.2% and *span* = 1.18%.

4.2. Macro-Scale: Crack Detection

The input parameters and material properties of the macro-scale model are listed in Table 3. In the macro-scale, we examine the model's capability of detecting a crack in the *ZY* plane introduced through the reduction of the element's electrical conductivity derived from the micro-scale analysis [14]. Two cracks (crack1 and crack2) of different length have been modeled at the same location; their characteristics are listed in Table 4.

Table 3. Input parameters and materials properties in the macro-scale model.

Parameter/Property	Value
Theoritical fiber volume fraction	55%
Transverse electrical conductivity variation	$[1.13 \times 10^{-4}, 3.47 \times 10^{-3}, 4.02 \times 10^{-3}, 5.77 \times 10^{-3}, 9.32 \times 10^{-3}]\,S/m$
Fiber's electrical conductivity at Z direction, σ^z_{cf}	$10^5\,S/m$
Electrical resistance of circuit elements	10^3 Ohm

Table 4. The characteristics of the cracks modeled.

	Position (*X*, *Z*) (mm)	Depth from Topurface (mm)	Length (μm)	Reduction in (%)
1	(6, 2)	−0.2	50	59
2	(6, 2)	−0.2	100	100

By comparing the current values taken from the circuit elements for the reference and the cracked model, we plot the results at the *XZ* plane in Figure 10. The obtained inhomogeneity of the current distribution is due to the inhomogeneous electrical properties of the model's elements.

The computed contours of *drop* at the *Y* direction due to the presence of crack1 and crack2 are plotted on the *XZ* plane in Figures 10 and 11, respectively. The results have been normalized by the current of the reference model. As shown, in both cases the location of the current drop at the *Y* direction matches the location of the crack. For crack1, given the crack spans through the width of the modified element, the ratio of the crack surface to the *XZ* surface of the model is $\frac{50\ \mu m \times 10\ \mu m}{10^4\ \mu m \times (2.5 \times 10^3)\ \mu m} = 0.00002 = 0.002\%$. Hence, the response is quite large compared to the modeled crack length. The reason for choosing the width of 0.2 mm (20% of the model's width) is because the meso-scale analyses have shown that smaller crack widths give very weak responses. In the case of crack2, the magnitude of electrical response increases by 643.3% for a 100% increase in crack length from 50 μm to 100 μm. This finding is an indication of the high sensitivity of the model to the crack length.

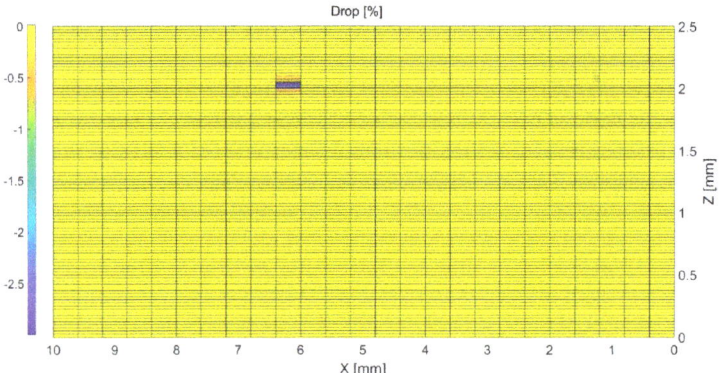

Figure 10. 2D plot of the drop at the *XZ* plane for crack1 (50 μm).

To further exploit the above findings, we note that the same technique does not have the same sensitivity in metallic materials since in that case the electric current from the cracked area is redistributed to the remaining material volume and the *drop* is not detectable. On the contrary, due to the very small volume of the CNT network compared to the volume of the composite material, any interruption of the network, which leads to the redistribution of electric current, is detectable.

Aiming to characterize crack1, we use the technique described in the previous section. Based on the computed values of *drop* and *span*, we delimit the area around the crack until the nodes where the electrical response is fully recovered (zero value). The dimension of the perturbation area is defined as the dimension in the direction of the denser mesh (*Z* direction). For crack1, we define an area of dimension of 250 μm and for this area we read *drop* = 3% and *span* = 1.1 (see Figure 3). The intersection of the *drop* and *span* intersection-curves shown in Figure 12 gives l_c/l_0 = 2.1, a crack length of 52.5 μm and d_c/l_0 = −1.88. The characterized values of crack length and crack depth are close to the actual values, something which validates the proposed crack sensing methodology.

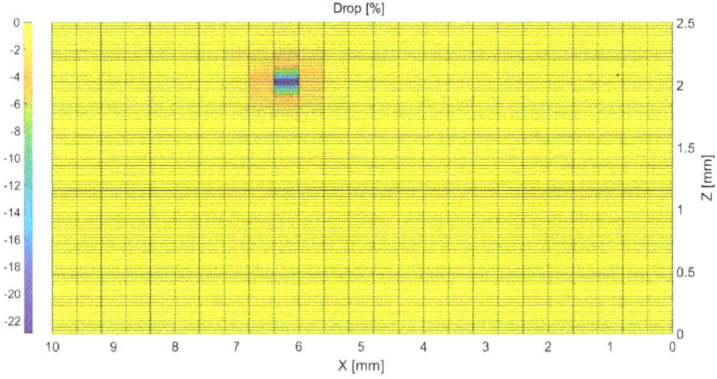

Figure 11. 2D plot of the drop at the *XZ* plane for crack1 (100 μm).

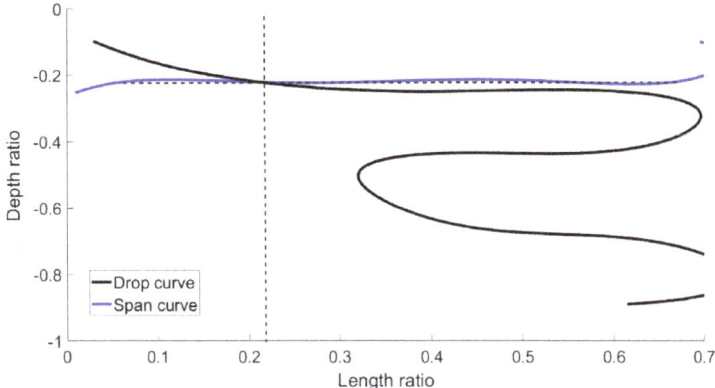

Figure 12. Intersection curves for *drop* = 3% and *span* = 1.1.

5. Conclusions

This is the second of a two-paper series describing a multi-scale modeling approach developed to simulate crack sensing in polymer fibrous composites by exploiting interruption of electrically conductive carbon nanotube (CNT) networks. In the present paper, the meso- and macro-scale analyses has been described. In the meso-scale, a crack sensing methodology has been developed by means of a parametric study which correlates the crack characteristics with the current drop characteristics. In the macro-scale, the final implementation of the crack sensing methodology has been made for cracks of different lengths. The numerical results show a large sensitivity of the current flow to the crack length. Moreover, the model has been proved capable of predicting both the length and the position of the cracks considered.

Based on the findings of the two papers, we can conclude that the proposed multi-scale electrical modeling approach is capable of simulating the electrical response of CNT/polymers and CNT/composites. The model in combination with the crack sensing methodology is also capable of characterizing cracks in these materials. The proposed modeling approach after being improved could be used for the virtual design and optimization of structural health monitoring systems based on conductive CNT networks in polymers, epoxy adhesives, and polymer fibrous composites.

Author Contributions: Conceptualization, K.T. and C.K.; methodology, K.T. and C.K.; software, C.K.; validation, K.T.; investigation, K.T. and C.K.; writing—original draft preparation, K.T. and C.K.; writing—review and editing, K.T.; supervision, K.T.; project administration, K.T.; funding acquisition, K.T.

Funding: Part of the work reported in this paper has received funding from the European Union's Horizon 2020 research and innovation programme ECO-COMPASS (Grant no. 690638).

Conflicts of Interest: The authors declare no conflict of interest.

References

1. Boller, C. Why SHM? A Motivation. In *NATO Lecture Series STO-MP-AVT-220 Structural Health Monitoring of Military Vehicles: STO Educational Notes*; NATO STO: Brussels, Belgium, 2014.
2. Duong, H.M.; Gong, F.; Liu, P.; Tran, T.Q. Advanced Fabrication and Properties of Aligned Carbon Nanotube Composites: Experiments and Modeling. Available online: https://www.intechopen.com/books/carbon-nanotubes-current-progress-of-their-polymer-composites/advanced-fabrication-and-properties-of-aligned-carbon-nanotube-composites-experiments-and-modeling (accessed on 5 October 2017).
3. Tran, T.Q.; Headrick, R.J.; Bengio, E.A.; Myo Myint, S.; Khoshnevis, H.; Jamali, V.; Duong, H.M.; Pasquali, M. Purification and Dissolution of Carbon Nanotube Fibers Spun from the Floating Catalyst Method. *Appl. Mater. Interfaces* **2017**, *9*, 37112–37119. [CrossRef] [PubMed]

4.   Tserpes, K.I.; Papanikos, P. Finite element modeling of single-walled carbon nanotubes. *Compos. Part B* **2005**, *36*, 468–477. [CrossRef]

5.   Papanikos, P.; Nikolopoulos, D.D.; Tserpes, K.I. Equivalent beams for carbon nanotubes. *Comput. Mater. Sci.* **2008**, *43*, 345–352. [CrossRef]

6.   Tserpes, K.I.; Papanikos, P.; Tsirkas, S.A. A progressive fracture model for carbon nanotubes. *Compos. Part. B* **2006**, *37*, 662–669. [CrossRef]

7.   Tserpes, K.I.; Papanikos, P. The effect of Stone-Wales defect on the tensile behavior and fracture of single-walled carbon nanotubes. *Compos. Struct.* **2007**, *79*, 581–589. [CrossRef]

8.   Kim, P.; Shi, L.; Majumdar, A.; Mc Euen, P.L. Thermal transport measurements of individual multiwalled nanotube. *Phys. Rev. Lett.* **2001**, *87*, 215502. [CrossRef] [PubMed]

9.   Ebbesen, T.W.; Lezec, H.J.; Hiura, H.; Bennett, J.W.; Ghaemi, H.F.; Thio, T. Electrical conductivity of individual carbon nanotubes. *Nature* **1996**, *382*, 54–56. [CrossRef]

10. Fiedler, B.; Gojny, F.H.; Wichmann, M.H.G.; Bauhofer, W.; Karl, S. Can carbon nanotubes be used to sense damage in composites? *Eur. J. Control.* **2004**, *29*, 81–94. [CrossRef]

11. Gallo, G.J.; Thostenson, E.T. Electrical characterization and modeling of carbon nanotube and carbon fiber self-sensing composites for enhanced sensing of microcracks. *Mater. Today Commun.* **2015**, *3*, 17–26. [CrossRef]

12. Aly, K.; Li, A.; Bradford, P.D. Strain sensing in composites using aligned carbon nanotube sheets embedded in the interlaminar region. *Compos. Part. A* **2016**, *90*, 536–548. [CrossRef]

13. Li, C.; Chou, T.W. Modeling of damage sensing in fiber composites using carbon nanotube networks. *Compos. Sci. Technol.* **2008**, *68*, 3373–3379. [CrossRef]

14. Kuronuma, Y.; Takeda, T.; Shindo, Y.; Narita, F.; Wei, Z. Electrical resistance-based strain sensing in carbon nanotube/polymer composites under tension: Analytical modeling and experiments. *Compos. Sci. Technol.* **2012**, *72*, 1678–1682. [CrossRef]

15. Tserpes, K.; Kora, Ch.   A multi-scale modeling approach for simulating crack sensing in polymer fibrous composites using electrically conductive carbon nanotube networks. Part II: Micro-scale analysis. *Comput. Mater. Sci.* **2018**, *154*, 530–537. [CrossRef]

16. *ANSYS User's Manual*, version 11; Swanson Analysis Systems: Pittsburgh, PA, USA, 2008.

Article

Effect of Ramie Fabric Chemical Treatments on the Physical Properties of Thermoset Polylactic Acid (PLA) Composites

Chunhong Wang [1,2,3,*]**, Zilong Ren** [1,2,3]**, Shan Li** [4] **and Xiaosu Yi** [5,6]

1   School of Textile, Tianjin Polytechnic University, Tianjin 300387, China; 13920566317@163.com
2   Key Laboratory of Advanced Textile Composite Materials, Tianjin Polytechnic University, Tianjin 300387, China
3   Key Laboratory of Hollow Fiber Membrane Material and Membrane Process of Ministry of Education, Tianjin Polytechnic University, Tianjin 300387, China
4   Chinatesta Textile Testing & Certification Services, Beijing 100025, China; lishan880131@126.com
5   Beijing Institute of Aeronautical Materials, Beijing 100095, China; xiaosu.yi@nottingham.edu.cn
6   Aviation Composite (Beijing) Science and Technology Co., Ltd., Beijing 101300, China
*   Correspondence: cn_wangch@163.com; Tel.: +86-022-8395-5608

Received: 13 July 2018; Accepted: 30 August 2018; Published: 2 September 2018

Abstract: Ramie fabric-reinforced thermoset polylactic acid (PLA) composites were prepared by using heat pressing technology. Fabrics were treated with alkali, silane, and alkali–silane respectively, expecting an improvement of the interface between the fabric and the matrix. Scanning electron microscopy (SEM) results indicated that after alkali treatment, impurities on the fiber surface were removed and its diameter became finer. After the silane, and alkali–silane treatments, the contact angles of the ramie fibers increased by 14.26%, and 33.12%, respectively. The contact angle of the alkali–silane treated fiber reached 76.41°; this is beneficial for the adhesion between ramie fiber and the PLA. The research revealed that the tensile strength of the fiber increased after the alkali and silane treatments. A slight decrease was noticed on the tensile strength of fibers treated with alkali–silane. After all, three chemical treatments were done, the flexure strength of the ramie fabric-reinforced PLA composites, improved in all cases. Among the three treatments, the alkali–silane treatment demonstrated the best result, as far as the flexure strength and modulus of the fabricated composites were concerned. On the other hand, water absorption of the related composites decreased by 23.70%, which might contribute to the closer contact between the ramie fiber and the matrix. The ramie fabric-reinforced PLA composites, prepared in this study, can meet the standard requirements of aircraft interior structures and have favorable application foreground.

Keywords: fabric; interface; physical properties; thermosetting resin

---

1. Introduction

In recent years, with the global environment and energy problems becoming increasingly prominent, the consciousness about ecological environmental protection is continuously being strengthened. Therefore, natural fibers become more popular to replace synthetic ones in the composites field [1–4]. However, most of the resins used to fabricate the composites are originated from oil, which is considered to not be environment-friendly. Moreover, oil is a non-renewable resource, and all countries face the problem of an oil shortage. Thus, biological resins have aroused wide interest as a substitute for oil-based resins because of renewable raw materials, such as soybean oil and rapeseed oil. Unlike petroleum oil, vegetable oils can be circularly obtained every year with no pollution of the environment [5].

Biological resins may include starch, polyhydroxybutyrate (PHB), thermoplastic polylactic acid, polybutylene succinate (PBS), and various other biological resins [6]. During the composite fabrication process, when using biological resins, injection, extrusion and compression molding processes may be employed [7]. Biological resin includes thermoset and thermoplastic resins. The toughness of thermoplastic resin is higher, but it cannot be dissolved easily in ethanol, acetone, and other general solvents. On the other hand, its performance is not considered to be stable [8]. Thermoset resin is a crosslinked polymer with a mesh structure, after heating with the curing agent included, it becomes insoluble and infusible. It is believed that thermoset matrix composites offer excellent thermal, mechanical and flexure performance and the composite fabrication process is simple and easy to control [9].

The raw materials of thermoset PLA are vegetable oils, such as soybean oil, linseed oil, and rapeseed oil. However, due to different synthetic processes, most of the thermoset PLA is nondegradable, such as acrylic epoxy soya oil resin (AESO)—researched by the University of Delaware (USA)—and bio-based resins derived from soybean oil in North Dakota State University.

Zhang et al. [10,11] prepared bio-matrix composites using phthalic anhydride modified AESO as a matrix and glass fibers as reinforcements for circuit boards. Adekunle [12] found that sisal/AESO/styrene composite has higher tensile strength, flexural strength, and flexural modulus. The effects of acidification, silane and peroxide modifications on the interfacial properties of ramie fiber/AESO composites were investigated by Lee [13]. The results showed that the interfacial shear strength of the composites was significantly improved after modifications.

In the study by Amiri [14], the mechanical properties of bio-based resin composites showed improvement when using alkaline-treated flax fibers as reinforcements, and their performances could meet the requirements of aircraft's interior structures. Amiri [15,16] studied the long-term creep behavior of flax/VE bio-based composites. It was found that the flax/VE composites could be considered to be thermorheologically complex materials. Amiri [17] investigated the effect of alkaline treatment of flax fiber as well as an addition of 1% acrylic resin to vinyl ester on mechanical properties and long-term creep behavior of flax/vinyl ester composites. Findings revealed that alkaline treatment was successful in increasing interlaminar shear, tensile and flexural strength of the composite but decreased the tensile and flexural modulus by 10%. Addition of acrylic resin to the vinyl ester resin improved all mechanical properties except the flexural modulus which was decreased by 5%. In a study by Taylor [18], a newly developed bio-based resin, epoxidized sucrose soyate (ESS), was combined with surface-treated flax fiber to produce novel bio-composites. The bio-composites properties could meet and exceed those of conventional pultruded members.

Distinguished from traditional studies, this paper used the bio-based thermoset polylactic acid (PLA), developed by Professor Jukka Seppala of Helsinki University of Technology (Finland). This PLA is completely degradable and is considered to be more environmental–friendlier, with a great potential for use [19–22]. It is based on a polylactic acid and its molecular formula is shown in Figure 1, where 'k', 'n', 'm', and 'p' refer to the polymerization degrees of different functional groups.

Ramie is a perennial herbaceous plant; its fiber is long, has great strength, as well as excellent thermal conductivity [23]. On the other hand, due to the strong polarity of the fiber, it offers a more hydrophilic, greater expansion, in a humid environment, and lower thermal stability [19,24].

Research studies have revealed that shortcomings exist for composites fabricated with natural fiber and the thermosetting resin. These fibers are more moisture sensitive, and the interface between the hydrophilic fiber and hydrophobic matrix is not closely combined [25–28]. It is well known that the interface plays a critical role in the mechanical performance of composites [29]. Therefore, improvement of the interface behavior is of extreme importance. The aim of this study is to investigate the effect of the chemical treatment on the properties of the ramie fiber as well as the produced composites.

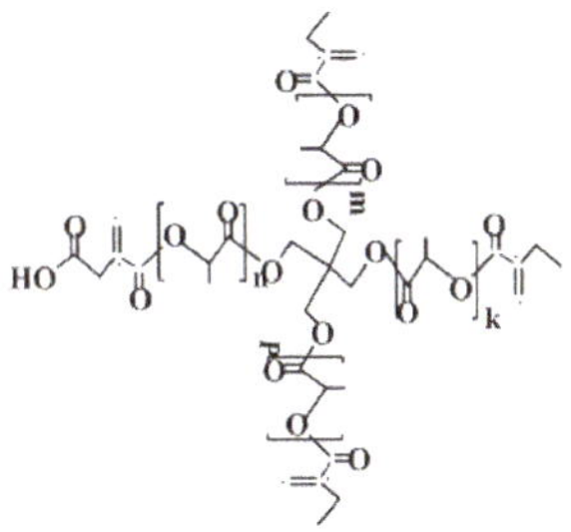

Figure 1. Polylactic acid (PLA) molecular formula.

In this investigation, composites with PLA were used as the resin and the ramie fabrics, with and without chemical treatment, were used as the reinforcement. They were prepared using the heat press technology. The morphology of the fibers and the composites was studied using scanning electron microscopy (SEM) and the mechanical properties were determined by Instron Model 3369. The purpose of this study was to explore whether ramie/PLA composites could be used in aerospace applications, to replace composites that use glass fiber and petroleum oil-based resins, so as to save energy and protect the environment.

2. Experimental Details

2.1. Materials

Ramie fabrics with the area density of 135.1 g/m² and thickness of 0.291 mm, were provided by Hunan Dongting Maye Company, China. The tensile strength of the warp and weft yarns was 44.81 MPa and 55.45 MPa, respectively. The thermosetting polylactic acid, with a flexure strength of 110.72 MPa, was provided by the Helsinki University of Technology, Finland. The curing agent, benzoyl peroxide dibutyl paste, was supplied by Tianjin Synthetic Materials Company, China. The release agent was produced by Beijing Carat Chemical Technology Company, China. Coupling agent KH550 (silane) was made by Nanjing Safer Silane Coupling Agent Factory, China. NaOH granules were given by Tianjin Wind Ship Chemical Technology Co. Ltd., Tianjin, China.

2.2. Surface Treatment

2.2.1. Alkali Treatment

The alkali solution concentration was 1 wt% (weight percentage compared to distilled water), the ramie fabric and alkali bath ratio was 1:20. After immersing the fabrics in the alkali liquid for 60 min, at a temperature of 60 °C, the fabrics were removed and washed to neutral, in warm water, and then cured in an electro-thermal blowing oven (Model DHG-9070A, Shanghai Scientific Instrument Co. Ltd., Shanghai, China) for 12 h. The temperature in the oven was kept at 80 °C during the drying process. All the process parameters of alkali treatment were derived from previous studies [30].

2.2.2. Silane Treatment

The silane (KH550) solution concentration was 3 wt% and the ramie fabric and silane bath ratio was 1:20. The fabric was soaked in the silane liquid for 2 h, at the room temperature, then was removed and cured for 12 h, in conditions identical to the alkali treatment. All the process parameters of silane treatment were derived from previous studies [30].

2.2.3. Combined Alkali–Silane Treatment

The fabric was first treated by the alkali treatment process, then the silane treatment process.

2.3. Fabrication of the Composites

Ramie fabric-reinforced PLA composites were prepared by using a hydraulic press (Model Y/TD71-45, Tianjin Hydraulic Machine Factory, Tianjin, China). Ramie fabric/PLA (*v/v*) was 60/40, crosslink agent/PLA (*w/w*) was 33/67, curing agent/PLA (*w/w*) was 1/99. The ramie fabric/PLA prepreg was first pressed for 10 min, at room temperature, with pressures up to 10 MPa. Then was heat pressed for another 60 min, at a temperature of 90 °C, with pressures up to 20 MPa, this made the percolation of PLA more uniform. Then it was pressed at pressures up to 20 MPa, at a temperature of 115 °C, for 2 min. The process continued with pressures up to 20 MPa, at a temperature of 130 °C, for 30 min, after which it was cooled down to the room temperature. At last the composites were removed from the press and post-cured at a temperature of 140 °C, for 120 min. In this study, time was recorded after the hydraulic press reached the set temperature, and the time required for the heating process was not included in the calculation.

2.4. Scanning Electron Microscope (SEM) Examination

The surface of ramie fibers and the composites were examined using a Model TM-1000 scanning electron microscope (HITICH, Japan), to reveal the morphological changes after the chemical treatments of the fabrics were carried out.

2.5. Contact Angle Test

Contact angle investigation was performed using the Model K100 contact angle instrument (Rruss Company, Emsdetten, Germany). During the measurement, the specific fiber length was 4 to 6 mm, fiber pull-out speed was 0.01 mm/s. The experiments were performed at a temperature of 20 °C and a relative humidity of 65%. The diameter of the ramie fiber was first measured by a microscope diameter analyzer (Model VHX-1000, Keyence International Trade Co. Ltd, Shanghai, China).

2.6. Tensile Strength Test

In order to explore the effect of chemical treatments on the fibers, tensile properties of ramie fibers were tested by a Model YG001A electronic power meter (Taicang Textile Instrument Factory, Suzhou, China). The test was carried out according to ASTM D3822-07 procedures. The clamping distance was 20 mm, and the stretching speed was 5 mm/min. Twenty samples in each case were included in the test.

2.7. Flexure Property

Composites were subjected to three-point bend test, in an Instron instrument (Model 3369, Instron Corporation, Boston, MA, USA), in accordance with the ASTM 790-03 Standard. The sample size was 60 mm × 12.5 mm × 3 mm. The span was 48 mm, loading gauge was 48 mm, and loading speed was 2 mm/min.

2.8. Water Absorption

The dimensions of the specimen were 60 mm × 12.5 mm. The specimens were dried at 60 °C for 24 h, then cooled down to room temperature, in a desiccator. After weight measurements were done, specimens were immersed in distilled water for 24 h, at the room temperature. Then water on the surface was wiped dry and weighed again. Five specimens in each case were included in the test.

3. Results and Discussion

3.1. Surface Morphology

Morphology of ramie fibers, both chemicals treated and untreated, are compared in Figure 2. It is understandable that some impurities which are rich in their content of hydroxyls, such as hemicelluloses, pectin, and lignin, existed on the surface of the untreated fibers. After the alkali treatment, the fiber surface became cleaner and some impurities were removed during the treatment [29]. However, the fiber surface also became rough [31] with an increase of the length/diameter ratio. This may contribute to the combination of the sodium hydroxide, in the hydroxyl, with the hydrogen bonding in the cellulose, resulting in a decrease of the active groups on the fiber surface. On the other hand, the hemicellulose, lignin, pectin, ester waxy, and biological oil on the fiber surface were reduced, causing an improvement of the surface purity and the reduction of the fiber diameter [5,25,32–35]. As for the silane treated fibers, impurities were noticed on the fiber surface. It was considered to be the membrane coupling agent [36,37]. Compared to the untreated fiber, no obvious difference was seen on the surface of the ramie fibers treated by alkali alone and those treated by the alkali–silane treatment. Similar results have been noticed by other researchers [21].

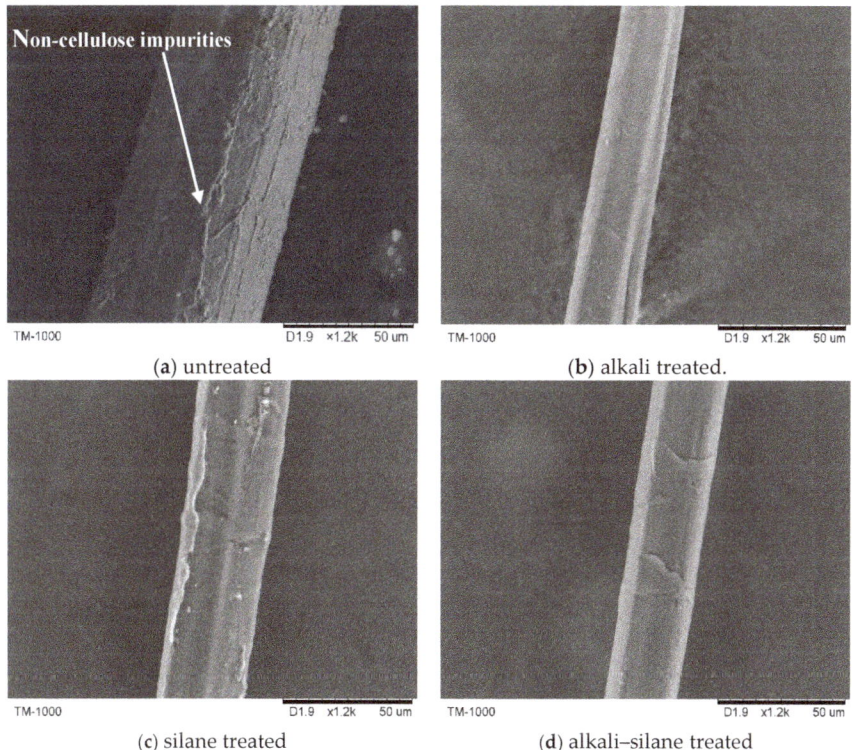

Figure 2. SEM photographs of ramie fibers with different modifications.

3.2. Contact Angle

Figure 3 shows the water contact angle of the treated and untreated ramie fibers. From the figure, one can see that after the alkali treatment, the water contact angle of ramie fiber increased from 57.39° to 58.76°, i.e., slightly increased by 2.38%. Alkali treatment could remove the lignin and other non-cellulosic substances so that the inner cellulose was exposed on the surface of ramie fiber. However, as lignin and cellulose are polar, alkali treatments had little effect on the polarity of the ramie fiber surface, and the contact angle more-or-less held the line.

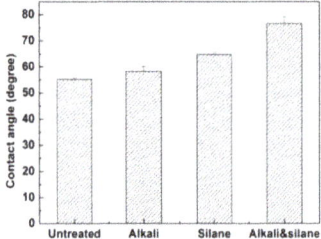

Figure 3. Water contact angle of ramie fibers with different modifications.

Contact angle on the ramie fiber increased by 14.26% after the silane treatment. The general formula of the organic sliane coupling agent is R_nSiX_{4-n}, the X group hydrolyzed into silicon alcohol (Figure 4), then the silicon hydroxyl reacted with the hydroxyl on the surface of the ramie fiber to form hydrogen bonds, the hydrogen then chemically bonded the coupling agent KH550, in combination with the ramie fiber (Figure 5). On the other hand, the silicon hydroxyls in the silicon alcohol reacted with each other to form hydrogen bonds, then the coupling agent KH550 on the ramie fiber surface changed into a film, through condensation. As Figure 2 shows, the impurities on the fiber surface were considered to be the membrane coupling agent. Due to the reaction between the X group and the ramie fiber, some organic groups (R) with a reactive functional were left on the fiber surface, which was beneficial for the increase of the contact angle. After the alkali–silane treatment, the contact angle increased by 33.12% and reached an angle, as large as 76.41°, this was the result of the double treatments.

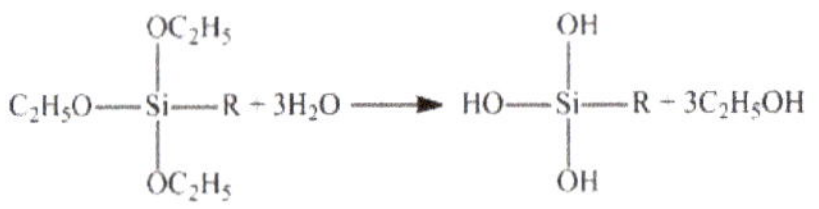

$$OC_2H_5 \qquad\qquad\qquad OH$$
$$| \qquad\qquad\qquad\qquad\qquad |$$
$$C_2H_5O——Si——R - 3H_2O \longrightarrow HO——Si——R - 3C_2H_5OH$$
$$| \qquad\qquad\qquad\qquad\qquad |$$
$$OC_2H_5 \qquad\qquad\qquad OH$$

Figure 4. Hydrolysis of coupling agent KH550.

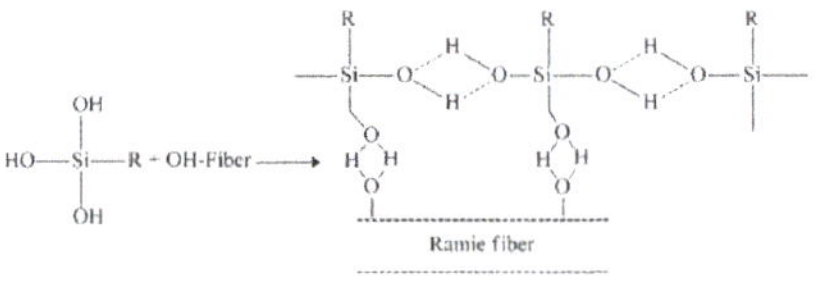

Figure 5. Reaction of coupling agent KH550 and the ramie fiber.

3.3. Single Fiber Tensile Properties

The tensile strength and modulus of the ramie fibers with different treatments are shown in Table 1. As seen in the table, the tensile strength of the fiber treated by the alkali solution received the highest value of 529.39 MPa, which increased by 32.53%, compared to that of the untreated fibers. This result is in line with previous research works [38]. The tensile strength of silane treated fibers was increased by 15.00%. Both of these increments may be a result of the reduction of impurities, causing an increase of the cellulose ratio on the fiber surface. The research of Cao et al. [39] supports this view. They found that after the alkali treatment, the content of the lignin and the pectin of flax fibers decreased by 37.04% and 53.85%, respectively, and the content of cellulose had increased. As has been well recognized, the crystal in the ramie fiber is composed of cellulose, the higher the degree of crystallinity, the higher the fiber tensile strength. Therefore, an increase in cellulose ratio would enhance the tensile strength of the ramie fiber. It was beneficial to enhance the mechanical performance of the resultant composites. A slight decrease was observed in the fiber tensile strength and the modulus, after the alkali–silane treatment. After alkali treatment, the impurities on the ramie fiber surface were removed. When the coupling agent treatment was carried out in the next step, the water molecules found it easier to enter into the ramie fiber, resulting in a swelling of the ramie fibers and a reduction of the bonding force between the cellulose chains. When subjected to an external force, the molecular chains of the cellulose were easily slipped, which would decrease the tensile strength of ramie fibers [40].

Table 1. Tensile strengths and modulus of ramie fibers with different modifications.

Treatment	Single Fiber Tensile Strength (MPa)		Single Fiber Tensile Modulus (GPa)	
	Mean	Standard Deviation	Mean	Standard Deviation
Untreated	399.47	178.74	18.30	7.88
Alkali treatment	529.40	233.41	17.40	9.09
Silane treatment	459.29	234.40	21.12	13.22
Alkali–silane treatment	386.03	221.23	15.62	5.47

3.4. Composites Flexure Property

The effect of the chemical treatments of the ramie fibers on its flexure strength and modulus of the composites is presented in Figure 6. It can be seen that the flexural properties of the composites increased after the alkali treatment, silane coupling treatment, and the alkali and silane treatment. Especially, the increase of the flexural strength and the modulus were highest for the alkali and silane treatment, and respectively increased by, 59.5% and 51.9%.

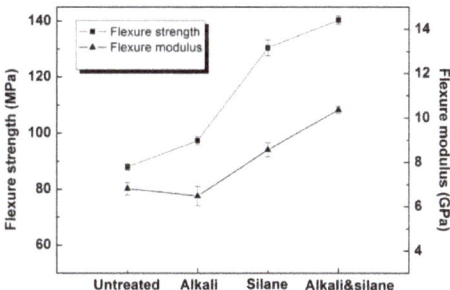

Figure 6. Effect of the different modifications on the flexure properties of ramie/PLA composites.

This improvement trend shows close consistencies with other research works [41]. The reason may contribute to the shrinkage of the ramie fabric after the alkali treatment [42]. The flexure strength and modulus of the composites, after the fiber was treated by silane, were increased by 48.25% and 25.70%, respectively. Due to the coupling agent KH550 film, which combined with the ramie fiber through hydrogen bonds, and the reactive functional organic groups (R) which were left on the fiber surface, organic groups (R), such as –C=C–, could have been involved in the curing reaction of PLA. Therefore, the bridge between the ramie fiber and the PLA resin was set up by the coupling agent KH550 (Figure 7). This was beneficial for the compatibility of the ramie fiber and PLA resin.

Composites received the highest flexure strength and modulus, with the values of 140.37 MPa and 10.36 GPa, respectively, when using the alkali–silane treated fabrics. This might be a result of the double treatment. It was understood that after the alkali–silane treatment, most of the non-cellulosic components in the fiber were removed [43,44]. The alkali treatment improved the hydrogen bonding effect and the silane treatment made the mechanical interlocking between the fiber and the PLA better.

Figure 8 illustrates the cross-section of the composites, when using the original and the alkali–silane treated fibers, respectively. An ideal fiber/matrix adhesion could be observed in Figure 8b; even more, the fibers in the matrix were well dispersed. It is understandable that a good dispersion and invasion of fibers into the matrix would result in ideal mechanical properties [45].

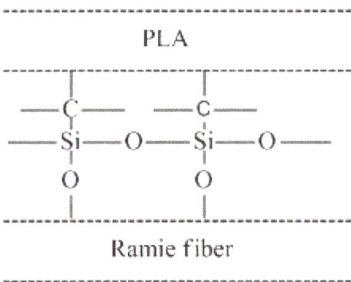

Figure 7. Reaction of the coupling agent KH550, the ramie fiber and the PLA.

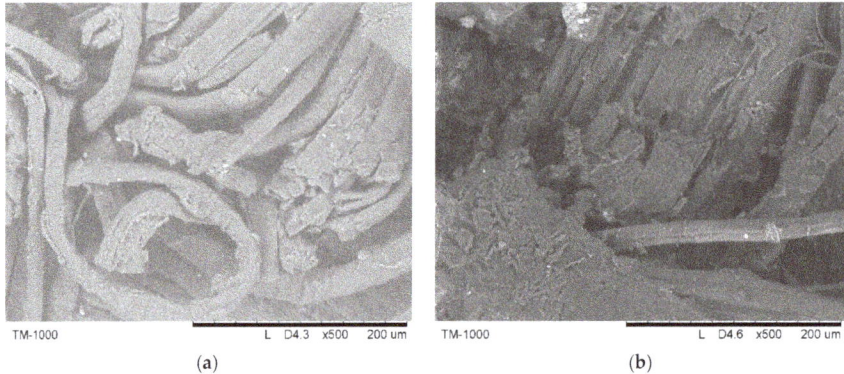

Figure 8. SEM fracture interface of ramie/PLA composites. (**a**) using original fabric and (**b**) using alkali–silane treated fabric.

3.5. Water Absorption Properties

Water absorption behavior of materials is critical to the dimensional stability of the final products.

The water absorption behavior of the composites, prepared in the research, is given in Figure 9. Compared to the synthetic fiber-reinforced composites, the water absorption rate for the ramie fiber-reinforced materials was much higher, this was true for all the specimens provided in this research. A possible explanation is the strong water imbibition of natural fibers [42]. Nevertheless, one could conclude that the fabric treating method played a major role, as far as the water absorption of the composites was concerned. As can be seen from Figure 9, the water absorption of ramie/PLA composites decreased to 14.97% after the alkali and the coupling agent treatment, and reduced by 23.70% compared to the untreated. The number of hydrophilic groups on the surface of ramie fiber, and the interfacial bonding between the fiber and the resin, were two important factors affecting the water absorption properties of composites. After the alkali and the coupling agent treatment, the hydroxyl groups (–OH) on the surface of ramie fiber reacted with silicon hydroxyl groups (Si–OH), produced by the hydrolysis of the coupling agent, and the number of hydrophilic groups, on the surface of ramie fiber, decrease. At the same time, the interfacial properties of the ramie/PLA composites were improved, and the internal structure was more compact, which might prohibit the penetration of the water molecules into the composites, so the water absorption of the composites was reduced.

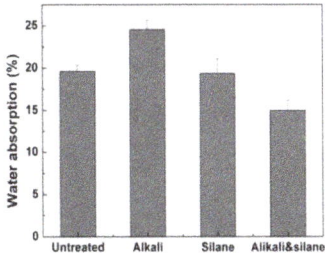

Figure 9. Effect of the different modifications on water absorption properties of the ramie/ PLA composites.

3.6. Application Performance Analysis

The application performance of ramie fabric-reinforced PLA composites is showed in Table 2. It can be seen that the ramie fabric-reinforced PLA composites, prepared in this study, were provided with excellent mechanical properties, and could be applied to an aircraft's interior structures, such as seats, console, floor and wall panels, luggage rack, etc.

Table 2. Application performance analysis of the ramie fabric/PLA composites.

Material	Flexural Strength/MPa
ramie fabric/PLA composites	140.37
Standard requirements for aircraft interior structures [46]	70

4. Conclusions

Ramie fabrics were treated by alkali, silane, and alkali–silane solutions, respectively. Then the fabrics were fabricated with thermoset polylactic acid (PLA) to prepare the composites. The research revealed that after the chemical treatments, the contact angle of the ramie fiber increased considerably. This is considered to be beneficial for the combination between the fiber and the matrix, during the composite processing.

Surface morphology observations indicated that the surface of the ramie fiber became much cleaner, after the alkali treatment, indicating that more impurities were removed during the chemical treatment. Meanwhile, after the alkali–silane treatment, the contact angle with water, increased, on the ramie fiber, and the tensile strength of the fiber improved significantly.

After molding, the flexure strength of the composites, when using fabrics treated by chemicals, was improved in each case. Among them, materials using the alkali–silane treated fabric received the highest flexure strength and modulus, of 140.37 MPa and 10.36 GPa, respectively.

Water absorption property of the composites was tested. The water absorption rate of the material, when using the alkali treated fabric, increased by 25.18%. No obvious change was noticed in the case of silane treatment. In the case of the alkali–silane treatment, the water adsorption of the material decreased by 23.70%. This may contribute to the closer fiber/matrix adhesion at the interface, in this case, as water molecules found it difficult to penetrate into the materials.

The ramie fabric-reinforced PLA composites, prepared in this study, can meet the standard requirements of an aircraft's interior structures and have a favorable application foreground.

Author Contributions: C.W. performed the design and guidance of experiments, data analysis; Z.R. and S.L. operated the experiments, and wrote the original draft preparation. X.Y. was responsible for the evaluation of application performance, and checked the accuracy of the data.

Funding: This research was funded by the National Natural Science Foundation of China, grant number 11802205, and the state key laboratory of automotive safety and energy of Tsinghua University, grant number KF1811.

Conflicts of Interest: The authors declare no conflicts of interest.

References

1. Scarponi, C.; P.izzinelli, C.S.; Sanchez-Saez, S. Impact load behaviour of resin transfer moulding (RTM) hemp fibre composite laminates. *J. Biobased Mater. Bioenergy* **2009**, *3*, 298–310. [CrossRef]
2. Khan, J.A.; Khan, M.A.; Islam, M.R. A study on mechnical, thermal and environmental degradation characteristics of N,N-Dimethylaniline treated jute fabric-reinforced polypropylene composites. *Fibers Polym.* **2014**, *15*, 823–830. [CrossRef]
3. Sarkar, B.K.; Ray, D. Effect of the defect concentration on the impact fatigue endurance of untreated and alkali treated jute-vinylester composites under normal and liquid nitrogen atmosphere. *Compos. Sci. Technol.* **2004**, *64*, 2213–2219. [CrossRef]
4. Ray, D.; Sarkar, B.K.; Rana, A.K. The mechanical properties of vinylester resin matrix composites reinforce with alkali-treated jute fibres. *Compos. Part A Appl. Sci. Manuf.* **2001**, *32*, 119–127. [CrossRef]
5. Wang, L.L.; He, L.P.; Tian, Y. Effects of surface modification on mechanical properties of sisal fiber reinforced polypropylene composites. *J. Mech. Eng. Mater.* **2008**, *32*, 58–61.
6. Chan, C.M.; Vandi, L.J.; Pratt, S. Composites of wood and biodegradable thermoplastics: A review. *Polym. Rev.* **2017**, *58*, 444–494. [CrossRef]
7. Faruk, O.; Bledzki, A.K.; Fink, H.P.; Sain, M. Biocomposites reinforced with natural fibers: 2000–2010. *Prog. Polym. Sci.* **2012**, *37*, 1552–1596. [CrossRef]
8. Zhang, N.; Zhang, J.J. Recent Progress in Study of Biomacromolecular Material Polylactide. *J. Hebei Polytecnic Univ. Nat. Sci. Ed.* **2010**, *32*, 116–120.
9. Helminen, A.O.; Korhonen, H.; Seppala, J.V. Structure modification and crosslinking of methacrylated polylactide oligomers. *J. Appl. Polym. Sci.* **2002**, *86*, 3616–3624. [CrossRef]
10. Zhan, M.; Wool, R.P. Biobased composite resins design for electronic materials. *J. Appl. Polym. Sci.* **2010**, *118*, 3274–3283. [CrossRef]
11. Lincoln, J.D.; Shapiro, A.A.; Earthman, J.C.; Saphores, J.D.M.; Ogunseitan, O.A. Design and evaluation of bio-based composites for printed circuit board application. *Compos. Part A Appl. Sci. Manuf.* **2013**, *47*, 22–30.
12. Adekunle, K.; Ghoreishi, R.; Ehsani, M.; Cho, S.; Skrifvars, M. Jute Fiber Reinforced Methacrylated Soybean Oil Based Thermoset Composites Prepared by Vacuum Injection Molding Technique. *J. Biobased Mater. Bioenergy* **2015**, *6*, 172–177. [CrossRef]
13. Lee, T.S.; Choi, H.Y.; Choi, H.N.; Kim, S.; Lee, S.G.; Yong, D.K. Effect of surface treatment of ramie fiber on the interfacial adhesion of ramie/acetylated epoxidized soybean oil (AESO) green composite. *J. Adhes. Sci. Technol.* **2013**, *27*, 1335–1347. [CrossRef]
14. Amiri, A.; Burkart, V.; Yu, A.; Webster, D.; Ulven, C. The potential of natural composite materials in structural design. *Sustain. Compos. Aerosp. Appl.* **2018**, 269–291. [CrossRef]
15. Amiri, A.; Ulven, C. Bio-based Composites as Thermorheologically Complex Materials. *Chall. Mech. Time Depend. Mater.* **2017**, *2*, 55–63.
16. Amiri, A.; Yu, A.; Webster, D.; Ulven, C. Bio-Based Resin Reinforced with Flax Fiber as Thermorheologically Complex Materials. *Polymers* **2016**, *8*, 153. [CrossRef]
17. Amiri, A.; Ulven, C.; Huo, S. Effect of Chemical Treatment of Flax Fiber and Resin Manipulation on Service Life of Their Composites Using Time-Temperature Superposition. *Polymers* **2015**, *7*, 1965–1978. [CrossRef]
18. Taylor, C.; Amiri, A.; Paramarta, A.; Ulven, C.; Webster, D. Development and weatherability of bio-based composites of structural quality using flax fiber and epoxidized sucrose soyate. *Mater. Des.* **2017**, *113*, 17–26. [CrossRef]
19. Oksman, K.; Skrifvars, M.; Selin, J.-F. Natural fibers as reinforcement in polylactic acid (PLA) composites. *Compos. Sci. Technol.* **2003**, *63*, 1317–1324. [CrossRef]

20. Sawpan, M.A.; Pickering, K.L.; Fernyhough, A. Improvement of mechanical performance of industrial hemp fibre reinforced polylactide biocomposites. *Compos. Part A Appl. Sci. Manuf.* **2011**, *42*, 310–319. [CrossRef]

21. Zhou, N.T.; Yu, B.; Sun, J. Influence of chemical treatments on the interfacial properties of ramie fiber reinforced poly(lactic acid)(PLA) composites. *J. Biobased Mater. Bioenergy* **2012**, *6*, 564–568. [CrossRef]

22. Aesson, D.; Skrifvars, M.; Lv, S. Preparation of nanocomposites from biobased thermoset resins by UV-curing. *Prog. Org. Coatings* **2010**, *67*, 281–286. [CrossRef]

23. Wang, H.G.; Xina, G.J.; Li, H. Durability study of a ramie-fiber reinforced henolic composite subjected to water immersion. *Fibers Polym.* **2014**, *15*, 1029–1034. [CrossRef]

24. Daniel, S.; Mustapha, A.; Christophe, P. Influence of hygrothermal ageing on the damage mechanisms of flax-fibre reinforced epoxy composites. *Compos. Part B Eng.* **2013**, *48*, 51–58.

25. Scarponi, C.; Schiavoni, E.; Sanchez-Saez, S. Polypropylene/hemp fabric reinforced composites: Manufacturing and mechanical behaviour. *J. Biobased Mater. Bioenergy* **2012**, *6*, 361–369. [CrossRef]

26. Asumani, O.M.L.; Reid, R.G.; Paskaramoorthy, R. The effects of alkali-silane treatment on the tensile and flexural properties of short fibre non-woven kenaf reinforced polypropylene composites. *Compos. Part A Appl. Sci. Manuf.* **2012**, *43*, 1431–1440. [CrossRef]

27. Goriarthi, B.K.; Suman, K.N.S.; Rao, N.M. Effect of fiber surface treatments on mechnical and abrasive wear performance of polylactide/jute composites. *Compos. Part A Appl. Sci. Manuf.* **2012**, *43*, 1800–1808. [CrossRef]

28. Kabir, M.M.; Wang, H.; Lau, K.T. Chemical treatments on plant-based natural fibre reinforced polymer composites: An overview. *Compos. Part B Eng.* **2012**, *43*, 2883–2892. [CrossRef]

29. Sawpan, M.A.; Pickering, K.L.; Fernyhough, A. Effect of fibre treatments on interfacial shear strength of hemp fibre reinforced polylactide and unsaturated polyester composites. *Compos. Part A Appl. Sci. Manuf.* **2011**, *49*, 1189–1196. [CrossRef]

30. Li, S.; Wang, C.H.; Yi, S.G.; Wang, R. Study on the preparation and flexural properties of ramie reinforced poly-lactic acid composites. *Fiber Reinf. Plast./Compos.* **2012**, *S1*, 59–62.

31. Nam, T.H.; Ogihara, S.; Tung, N.H. Effect of alkali treatment on interfacial and mechanical properties of coir fiber reinforced poly(butylen succinate) biodegradable composites. *Compos. Part B Eng.* **2011**, *42*, 1648–1656. [CrossRef]

32. John, M.J.; Anandjiwala, R.D. Recent developments in chemical modification and characterization of natural fibre-reinforced composites. *Polym. Compos.* **2008**, *29*, 187–207. [CrossRef]

33. Li, X.; Tabil, L.G.; Panigrahi, S. Chemical treatment of natural fibre for use in natural fibre-reinforced composites: A review. *Polym. Environ.* **2007**, *15*, 25–33. [CrossRef]

34. Mwaikambo, L.Y.; Tucker, N.; Clark, A.J. Mechanical properties of hemp fibre reinforced euphorbia composites. *Macromol. Mater. Eng.* **2007**, *292*, 993–1000. [CrossRef]

35. Ray, D.; Sarkar, B.K.; Rana, A.K. Effect of alkali treated jute fibres on composite properties. *Bull. Mater. Sci.* **2001**, *24*, 129–135. [CrossRef]

36. Wang, W.H.; Lu, G.J. The silane coupling agent treatment of basalt fibers reinforced wood-plastic composite. *Acta Mater. Compos. Sin.* **2013**, *30*, 315–320.

37. Guo, Z.F.; Zhong, Z.L. The effect of silane coupling agent treatment on tensile property of basalt fiber fabric. *Shanghai Text. Sci. Technol.* **2012**, *42*, 25–27.

38. Prasad, S.V.; Pavithran, C.; Rohatgi, P.K. Alkali treatment of coir fibers for coir–polyester composites. *J. Mater. Sci.* **1983**, *18*, 1443–1454. [CrossRef]

39. Cao, Y.; Wu, L.L.; Yu, J.Y. Influence of alkali treatment time on index of flax spinnability. *Text. Aux.* **2009**, *26*, 43–46.

40. Arbelaiz, A.; Cantero, G.; Fernandez, B.; Mondragon, I.; Gañán, P.; Kenny, J.M. Flax fiber surface modifications:effects on fiber physico mechanical and flax polypropylene interface properties. *Polym. Compos.* **2005**, *26*, 324–332. [CrossRef]

41. Moyeenuddin, A.S.; Kim, L.P.; Fernyhough, A. Flexural properties of hemp fibre reinforced polylactide and unsaturated polyester composites. *Compos. Part. A Appl. Sci. Manuf.* **2012**, *43*, 519–526.

42. Sahoo, S.; Nakai, A.; Kotaki, M. Mechanical properties and durability of jute reinforced thermosetting composites. *J. Biobased Mater. Bioenergy* **2007**, *1*, 427–436. [CrossRef]

43. Islam, M.S.; Pickering, K.L.; Foreman, N.J. Influence of alkali treatment on the interfacial and physico-mechanical properties of industrial hemp fibre reinforced polylactic acid composites. *Compos. Part A Appl. Sci. Manuf.* **2010**, *41*, 596–603. [CrossRef]

44. Kayode, A.; Riam, G.; Mehdi, E. Jute fiber reinforced methacrylated soybean oil based thermoset composites prepared by vacuum injection molding technique. *J. Biobased Mater. Bioenergy* **2012**, *6*, 172–177.

45. Lu, T.J.; Jiang, M.; Hui, D.; Wang, Z.; Zhou, Z. Effect of surface modification of bamboo cellulose fibers on mechanical properties of cellulose/epoxy composites. *Compos. Part. B Eng.* **2013**, *51*, 28–34. [CrossRef]

46. Editorial Board. *China Aerospace Materials Handbook*; Standards Press of China: Beijing, China, 2002.

Article

Life Cycle Assessment of Ramie Fiber Used for FRPs

Shaoce Dong [1], Guijun Xian [1,*] and Xiao-Su Yi [2,3]

[1]   School of Civil Engineering, Harbin Institute of Technology (HIT), 73 Hanghe Road, Nangang District, Harbin 150090, China; 18846135354@163.com
[2]   Faculty of Science & Engineering, University of Nottingham Ningbo China (UNNC), Ningbo 315000, China; xiaosu.yi@nottingham.edu.cn
[3]   National Key Laboratory of Advanced Composites, AVIC Composite Technology Center, AVIC Composite Corporation Ltd. (ACC), Beijing 101300, China
*   Correspondence: gjxian@hit.edu.cn; Tel.: +86-0451-8628-3120

Received: 15 June 2018; Accepted: 31 July 2018; Published: 3 August 2018

Abstract: With the depletion of natural resources and the deterioration of environment, natural fiber based biomaterials are attracting more and more attentions. Natural fibers are considered to be renewable, biodegradable, and ecofriendly, and have been applied to be used as alternative reinforcements to traditional glass fibers for polymer based composites (GFRP). Natural fiber reinforced polymer (NFRP) composites have been found to be manufactured as secondary structures or interior parts of aircrafts or automobiles. In this paper, a cradle-to-gate life cycle assessment (LCA) study was performed to demonstrate the possible advantages of ramie fiber on environmental impacts and to provide fundamental data for the further assessment of ramie fiber reinforced polymers (RFRP) and its structures. By collecting the material inventories of the production process of ramie fiber, the environmental impacts of ramie fiber (characterized by eight main impact categories, which are climate change, terrestrial acidification, freshwater eutrophication, human toxicity potential, ozone depletion, photochemical oxidant creation, freshwater ecotoxicity, and fossil depletion) were calculated and compared with that of glass fiber. Found if spinning process is ignored within the production of the ramie fiber, ramie fiber exhibits better ozone depletion and they have almost the same values of climate change and terrestrial acidification in terms of glass fiber. However, if the spinning process is included, ramie fiber only performs better in terms of ozone depletion. And degumming and carding and spinning processes are the processes that cause more pollution.

Keywords: ramie fiber; life cycle assessment; glass fibre; environmental impacts

---

1. Introduction

Fiber reinforced polymers (FRPs) have been widely used in modern industries because of their numerous advantages, like high specific strength and modulus, excellent fatigue performances, resistance to corrosion, and so on. The areas in which FRPs have been applied include aircraft, automotive, marine, sporting goods, infrastructure, and so on [1].

The fiber reinforcements used for FRPs are generally carbon- or glass-fibers. The production of such traditional fibers will bring in high emissions of greenhouse gases and the depletion of raw material natural resources (e.g., fossil fuel). Natural fibers are introduced and used in secondary or decorative structures considering their mechanical properties, because they are considered to be raw materials available, biodegradable, eco-friendly. Recently, natural fibers (e.g., ramie fiber, flax fiber, etc.) have been developed and utilized as a substitute for traditional glass fiber, to serve as fiber reinforcements of FRPs [2–4].

The life cycle assessment (LCA) of FRPs has received extensive attention for a better understanding of the environmental problems of FRPs and its structures. For example, regarding the traditional carbon FRPs (CFRPs) and glass FRPs (GFRPs), Foraboschi [5,6] studied the environmental problems of using FRP to strengthen masonry and reinforced concrete (RC) structures. For natural fiber reinforced polymers (NFRPS), there are many LCA studies on various natural fibers. Le Duigou et al. [7], and Summerscales and Dissanayake [8] have studied the environmental problems of flax fiber used in FRPs. Also, a comparative life cycle assessment (LCA) study was done between two bus body components made of hemp fiber reinforced polymers (HFRP) and GFRP [3]. It was found that the hemp-based composites show a lower environmental impact. Furthermore, LCA studies of jute fiber and sisal fiber have also been found [9,10].

In this article, the LCA of another widely used natural fiber, ramie fiber, was studied, because it is one of the best fibers that has better mechanical properties [11] in natural fibers and it can be used in ramie fiber reinforced polymers (RFRP). To the best of our knowledge, research about the environmental problems of ramie fiber is still limited.

Ramie, known as Chinese grass, is a perennial herbaceous plant of the Urticaceae family and it is mainly cultivated in Asia [12]. China contributes more than 90% of the cultivation of ramie in the world [13]. In Chinese major production areas, ramie can be harvested three times in one year [12]. As ramie fibers are abundant and can meet the requirements of FRP manufacturing, in the European Union's Horizon 2020 research and innovation programme between China and European (Grant No., 690638), ramie fiber was selected to manufacture some of the secondary structures of airports.

In the present study, a cradle-to-gate LCA was performed for the ramie fiber in order to demonstrate the possible advantages of ramie fiber on the environmental impacts and to provide fundamental data for further the assessment of RFRP and its structures. The inventories of the processes, including ramie cultivation, harvesting, peeling, transportation, degumming, and carding and spinning, were collected and calculated. Then, eight environmental categories, namely, climate change, terrestrial acidification, freshwater eutrophication, human toxicity potential, ozone depletion, photochemical oxidant creation, freshwater ecotoxicity, and fossil depletion, were used to characterize the environmental impacts of the ramie fiber. Finally, the environmental impacts of the ramie fiber were compared with that of glass fiber, based on the same mass. A professional LCA software, GaBi, was used to perform the LCA study [14]. The results show that if the spinning process is ignored within the production of the ramie fiber, the ramie fiber exhibits better ozone depletion and they have almost the same values on categories like climate change and terrestrial acidification, in terms of glass fibre. However, if the spinning process is included, the ramie fiber performs only better in term of the ozone depletion. The degumming, and carding and spinning processes are the processes that cause more pollution.

2. Methodology

Quantitative LCA was the research methodology that was used in the present study to quantify the environmental performance of the ramie fiber. LCA is an internationally accepted method to assess the environmental impacts of a product or product system. With the collected and/or calculated inputs and outputs of the whole life cycle of a product or product system, LCA can be performed according to some standards (ISO 14040 and ISO 14044 [15,16]), in order to evaluate the related environmental impacts quantitatively. A complete LCA study includes four stages, goal and scope definition, life cycle inventory analysis (LCI), life cycle impact assessment (LCIA), and life cycle interpretation [15]. The following explicitly explains the calculation and/or estimation methods used.

2.1. Goal and Scope Definition

The goal of an LCA study is related to its intended application and audience, while the scope of an LCA is related to the studied system. The goal of the present study is to get the fundamental data regarding the environmental impacts of ramie fibers. The scope of this study includes ramie cultivation and harvesting, peeling, transportation, degumming, carding and spinning (shown in Figure 1).

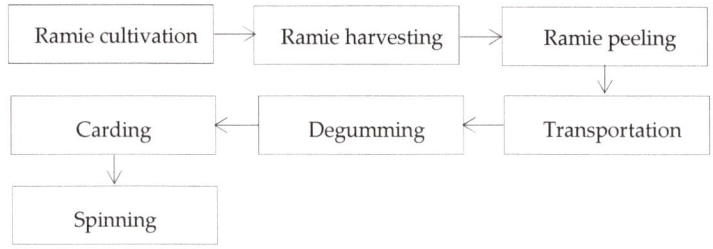

Figure 1. Production processes of ramie fiber.

2.2. Functional Unit

The functional unit is specified to provide a reference to which the inputs and outputs are connected. As the specific modulus of the ramie fiber and glass fiber are both 29×10^5 m [17], the functional unit was defined as 1000 kg of ramie fiber, which can offer the same stiffness as glass fiber on a weight-for-weight basis. However, where the composites are concerned, the lower density of the ramie fiber would lead to an increasing use of polymers in the RFRP rather than in the GFRP, if the volume content of polymers in both FRPs is the same.

2.3. Allocation Principle

Allocation is very important during the performance of LCA, and different allocation methods and a different set up of the primary products would affect the final environmental burdens of the product or product system studied [8,18]. In the present study, the allocation was performed twice based on the mass ratio of primary products to the by-products. At first, the allocation was done in the ramie cultivation, harvesting, and peeling process, where raw ramie fiber was considered as the primary product and the ramie leaves, ramie stalk, and ramie bark (excluding raw ramie fiber) were considered as by-products. Also, in the carding and spinning process, allocation was performed again and long fibers were considered as a primary product, while short fibers and shives were considered as by-products.

The environmental impacts are characterized by eight categories, which are climate change, freshwater eutrophication, ozone depletion, human toxicity, photochemical oxidant creation and terrestrial acidification, freshwater ecotoxicity, and fossil depletion, and were allocated based on the principle stated above.

2.4. Life Cycle Inventory Analysis

The life cycle inventory analysis is the second step to complete a LCA. In this step, all of the inputs and outputs related to the processes within the scope need to be collected. According to the scope of the study, the inputs and outputs of the ramie cultivation, harvesting, peeling, transportation, degumming, and spinning processes need to be collected. The data used in this article comes from published papers, theses, websites, and also a ramie textile mill, named Hunan Huansheng Dongting Maye Company Limited, which is located in Yueyang city, Hunan province, China.

2.4.1. Inputs and Outputs of Ramie Cultivation, Harvesting, and Peeling

As for the inputs of ramie cultivation, there is no need for seeding after each harvesting, because ramie is a perennial herbaceous plant and cuttage propagation is the main method for planting ramie. Averagely 12.8 kg N, 3.2 kg P_2O_5, and 16.5 kg K_2O are needed for producing 100 kg of raw ramie fiber [19,20]. To decrease the weeds, 7.5 kg diuron together with 40 kg water per hectare [19] are used, and 0.42 kg cyhalothrin together with 675 kg water per hectare are assumed to be used for reducing the number of pesticides. It is assumed that a chisel plough was used in winter for weeding once a year, and the amount of diesel consumed was set as 8.8 L/ha [21]. The amount of diesel consumed by agricultural equipment for spraying agrochemicals was set as 1 L/ha [21]. It is assumed that no irrigation was applied during the growth of ramie fiber.

Concerning the outputs of ramie cultivation, 3142.9 kg ramie leaves, 2357.1 kg ramie stalk, together with 1642.9 kg ramie bark (including raw ramie fiber) will be obtained when 1000 kg of raw ramie fiber is obtained [19]. An average value (2337 kg per year per hectare) of raw ramie fiber yield was set according to the real situation in Hunan province, China, because Hunan province (central area of China) is the main ramie production area in China and the cooperative ramie textile mill is located in Hunan province. Table 1 shows the annual raw ramie yield of Hunan province from 2000 to 2016, with missed data of 2009–2014, which is reported by the ministry of agriculture, China [22]. Moreover, the application of fertilizers will cause emissions of nitrogen gaseous fluxes and nitrogen leaching, which can be estimated according to Intergovernmental Panel on Climate Change (IPCC) tier 1 method [23]. Also, the emission factors of cyhalothrin to air, water, and soil are set as 0.01, 0.005, and 0.013 [24]. Phosphorus leaching was estimated according to the emission factor 0.01 [25]. The emissions of producing fertilizers, pesticide, herbicide, and water were acquired from GaBi databases and the pesticide and herbicide used in GaBi software to compute the environmental impacts were an unspecific type.

Table 1. Yields of ramie fiber of 2000–2016 with missed data of 2009–2014 in Hunan province.

Year	2000	2001	2002	2003	2004	2005	2006	2007	2008	2015	2016
Yield/kg	1964	2002	2359	2304	2443	2541	2516	2494	2502	2297	2285

The emissions from producing and the combustion of diesel are the main causes for environmental pollution in the ramie harvesting process. A 4 LMZ 160 crawler-type ramie harvester with 25.7 kW auxiliary power was used for harvesting the ramie fiber and its average harvesting rate is 0.15 hectare per hour [26]. When the consumption of diesel used for harvesting was computed, the specific fuel consumption was set as 0.2448 kg/kW·h, according to the data sheet of T35 type of diesel engine (whose power is 24 kW) of Changchai Company Limited [27]. As for diesel combustion, the emission factors of carbon monoxide, carbon dioxide, nitrogen oxides, sulfur dioxide, and non-methane volatile organic compounds (NMVOC) were set as 0.0291 g/kg, 3.04 g/kg, 0.0571 g/kg, 0.00415 g/kg, and 0.00916 g/kg, respectively [24].

As for the peeling process, generating electricity is the main reason accounting for emissions to the natural environment. A 6BZ-400 model ramie barker was assumed to peel the bark from the ramie stem, and its auxiliary power and peeling rate are 2.8 kW and 10 kg/h. The emissions of generating electricity were taken from GaBi.

An allocation was performed during this process based on the mass of raw ramie fiber, ramie leaves, ramie stalk, and ramie barks (excluding raw ramie fiber) and 14% of the environmental burdens were assigned to raw ramie fiber. The allocation was performed with the help of GaBi software. The biogenetic carbon storage of the ramie fiber was also considered in this work. The amount of carbon stored by the natural fiber depends on its cellulose content [28]. The degummed ramie fiber mainly consists of cellulose, and according to the literature [28], and when 1 kg of natural fiber is produced, 1.5 kg CO_2 would be absorbed.

Table 2 summarizes the inputs and outputs of producing 1000 kg of raw ramie fiber.

Table 2. Inputs and output of producing 1000 kg of raw ramie fiber.

Input	Quantity	Output	Quantity
Ammonium nitrate (33.5% N)	382.1 kg	Raw ramie fiber	1000 kg
Triple superphosphate (46% P_2O_5)	69.6 kg	Ramie bark (excluding raw ramie fiber)	642.9 kg
Potassium chloride (60% K_2O)	275 kg	Ramie leaf	3142.9 kg
Diuron	9.6 kg	Ramie stalk	2357.1 kg
Cyhalothrin	0.5 kg	Carbon monoxide	1.7 g
Water	917.8 kg	Carbon dioxide	176.5 g
Electricity	280 kW·h	Nitrous oxide	5.4 kg
Diesel	58 kg	Nitrogen oxides	3.3 g
Sequestered atmospheric CO_2	1500 kg	Sulfur dioxide	0.2 g
		Emission of cyhalothrin to air	5 g
		Emission of cyhalothrin to water	2.5 g
		Emission of cyhalothrin to soil	6.5 g
		Phosphate to water	0.14 kg
		NMVOC	0.5 g

This table does not include the inputs of the ramie rootstock and soil components consumed, which are about four tons. NMVOC—non-methane volatile organic compounds.

2.4.2. Inputs and Outputs of Transportation

The GaBi software contains almost all of the transport modes. In the present paper, the transportation process was modeled with GaBi. The transportation distance between the ramie field and the manufacturing factory is assumed as 50 km. A Euro 5 diesel truck with a 3.3 ton payload capacity was set to be the transportation method, considering the yield of the ramie fiber. Although the distance from field to factory is difficult to determine precisely, the environmental impacts of this process contribute very little to the overall environmental impacts, which can be seen from later parts.

2.4.3. Inputs and Outputs of Degumming Process

Because there are gummy matters (20–40% of the mass of raw ramie fiber), mainly composed of hemicellulose and pectin in raw ramie fibers [29], the raw ramie fiber cannot be used directly in FRPs until they are degummed. Degumming methods mainly include chemical, enzymatic, and microbial methods [30], and in the present study, enzymes together with chemicals (shown in Figure 2) were used to get the degummed fiber in this study.

The inputs of the degumming process were the chemicals (shown in Table 3) used to dissolve and remove the gummy matters, which were collected from the cooperative company mentioned above. It is known that, when 100 kg of raw ramie fibers were degummed, 60 kg degummed ramie fiber would be obtained, and 11 kg sulphuric acid, 20.25 kg sodium hydroxide, 2.4 kg hypochlorous, and 1.4 kg degumming agent would be consumed. The emissions for producing the chemicals used in this process were taken from the GaBi databases, except for degumming agent because of commercial confidentiality. As the dosage of the degumming agent is usually very small, the environmental impacts of producing and using the degumming agent have not been taken into consideration in this study. The environmental impacts of producing enzymes were not included in this paper, because of a lack of reliable data. A lack of data regarding the environmental impacts of hypochlorous acid, when the GaBi model was built, led to the use of calcium hypochlorite in order to produce hypochlorous acid, according to a chemical reaction among calcium hypochlorite, carbon dioxide, and water.

The outputs of the degumming process were mainly wastewater and it mainly comes from the pickling, boiling, washing, and bleaching processes. The wastewater from ramie degumming process usually has high pH value, high contents of chemical oxygen demand (COD), and biochemical oxygen demand (BOD). The wastewater cannot be discharged directly until it has been well treated. In this study, the wastewater treatment method was set based on the literature [31], in which a acidifying hydrolysis/biological oxidation/photochemical process method was utilized in a sewage treatment plant to make the wastewater meet the requirements of the national sewage discharge standard in China. To run the whole treatment process, electricity and chlorine need to be supplied constantly and the environmental impacts of producing electricity and chlorine were also within the system boundary in the present study. After treatment, the contents of COD, BOD, and sulfide were set as 300 mg/L, 100 mg/L according to the national wastewater discharge standard in China.

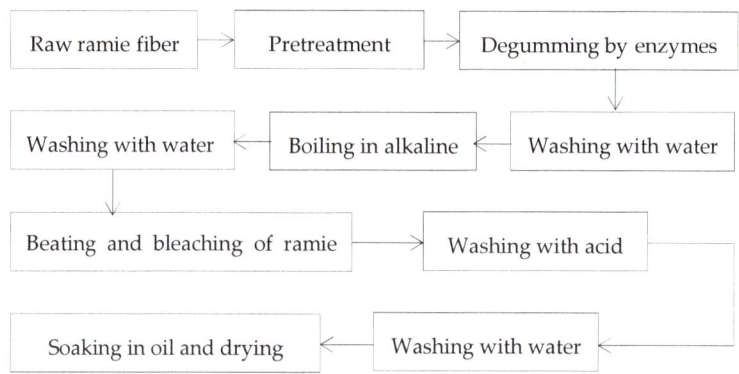

Figure 2. Typical enzymatic degumming processes of ramie fiber.

Table 3 shows the inputs and outputs for producing 1000 kg of degummed ramie fiber and the emissions of producing water, electricity, chlorine, sulphuric acid, and sodium hydroxide were taken from GaBi software.

Table 3. Inputs and outputs of producing 1000 kg of degummed ramie fiber. COD—chemical oxygen demand; BOD—biochemical oxygen demand.

Input	Quantity	Output	Quantity
Water	733 tons	Degummed ramie fiber	1000 kg
Electricity	1475.52 kW·h	Pectin, lignin, and hemicellulose	666.7 kg
Chlorine	8.8 kg	Wastewater (treated)	733 tons
Sulphuric acid	183.3 kg	COD	300 mg/L
Sodium hydroxide	337.5 kg	BOD	100 mg/L
Hypochlorous acid	40.0 kg		
Ramie degumming agent	23.3 kg		
Steam from biomass	16.7 tons		
Raw ramie fiber	1666.7 kg		

2.4.4. Inputs and Outputs of Carding and Spinning Process

Degummed ramie fibers could be spun to get yarns, which were used to for the comparison to glass filaments. Before spinning, carding is necessary to get long degummed ramie fibers suitable for spinning, and short fibers and shives that could be used in other ways like building particle boards or animal beddings. Therefore, the allocation was performed again in this stage, based on the mass of long fibers and others. Based on the data from the cooperative factory, only 30% of the degummed fibers will be formed into yarns. The electricity consumption for the carding and spinning are based on the data for the flax fiber carding and spinning [21], because of no reliable data are available for ramie carding and spinning. Table 4 shows the inputs and outputs of the ramie carding and spinning process.

Table 4. Inputs and outputs of producing 1000 kg ramie yarns.

Input	Quantity	Output	Quantity
Electricity	7260 kW·h	Ramie yarns	1000 kg
Degummed ramie fiber	3333.3 kg	Short fibers and shives	2333.3 kg

2.5. Life Cycle Impact Assessment (LCIA)

In the LCIA stage, the emissions and resource extraction will be translated into several environmental impact scores by multiplying the corresponding characterization factors. There are mainly two types of characterization factors, which are at a midpoint level and endpoint level, respectively.

In this paper, the ReCiPe method was chosen to model the environmental impacts of different fibers with the help of GaBi software. The environmental impact categories used to characterize the overall environmental performance of every fiber are climate change, terrestrial acidification, freshwater eutrophication, ozone depletion, human toxicity, photochemical oxidants creation, freshwater eco-toxicity, and fossil depletion.

The inputs and outputs of the processes involved in the scope of this study were used to build models in GaBi software.

2.6. Interpretation

Interpretation was the final step of an LCA study, and by interpretating the results from LCIA stage, the processes that have a high potential for pollution would be identified, and some useful conclusions and recommendations can be made.

3. Results and Discussions

Using the inputs and outputs from the LCI stage, the ReCiPe method and the eight main environmental categories to characterize the environmental performance of the ramie fiber, the LCIA results are shown in Table 5 and Figure 3.

Table 5. Life cycle impact assessment (LCIA) results of one ton ramie yarns, ramie fiber (after carding), and glass fibers.

Environmental Impact Category	Unit	Glass Fibers	Ramie Fiber (after Carding)	Ramie Yarns
Climate change	kg CO_2-Equiv.	1740	1770	3790
Terrestrial acidification	kg SO_2 eq	10.3	10.9	13.4
Freshwater eutrophication	kg P eq.	5.25×10^{-3}	86.9×10^{-3}	87.1×10^{-3}
Ozone depletion	kg CFC-11 eq.	483×10^{-10}	7.88×10^{-10}	8.26×10^{-10}
Human toxicity potential	kg 1,4-DB eq.	20.8	147	175
Photochemical oxidant formation	kg NMVOC	5.26	11.9	14.4
Freshwater ecotoxicity	kg 1,4-DB eq.	0.461	4.12	4.15
Fossil depletion	kg oil eq.	578	832	1.34×10^3

As can be seen from Table 5 and Figure 3, when the environmental burdens of the ramie yarns were compared to that of the glass fibers, the ramie yarns only performed better in terms of ozone depletion, with a reduction rate of 98.29%. However, as for the other categories, the glass fibers are more eco-friendly compared with the ramie yarns. When the environmental performance of the ramie fibers (after carding and without spinning) are compared to that of the glass fibers, the ramie fibers (after carding and without spinning) perform better on ozone depletion, with a reduction rate of 98.37%, and it has almost the same values for climate change and terrestrial acidification compared to glass fiber.

Concerning identifying the processes that contribute the most to the overall environment pollution, as can be seen Table 6 and Figure 4, the degumming process and carding and spinning process are the main causes for massive environmental pollution. The percentages of the degumming process and spinning and carding process that contributes to climate change, terrestrial acidification, freshwater eutrophication, ozone depletion, human toxicity, photochemical oxidant formation, freshwater ecotoxicity, and fossil depletion is 65.2% and 58.3%, 76.9% and 20.4%, 74.2% and 0.28%, 94.3% and 5.05%, 81.4% and 17.5%, 79.2% and 18.8%, 68.7% and 0.95%, and 53.7% and 41.1%, respectively. The cultivation, harvesting, and peeling processes mainly contribute to freshwater eutrophication, ozone depletion, and freshwater ecotoxicity, and the negative value caused by these processes are due to the absorption of carbon dioxide. The transportation process contributes little to all eight of these categories, compared to other processes.

Table 6. Values of the eight categories contributed by different production processes of ramie yarns.

Environmental Impact Category	Unit	Transportation	Cultivation, Harvesting and Peeling	Carding and Spinning	Degumming
Climate change	kg CO_2-Equiv.	31	−921	2210	2400
Terrestrial acidification	kg SO_2 eq.	0.024	0.346	2.74	4.94
Freshwater eutrophication	kg P eq.	5.9×10^{-5}	0.0125	24.1×10^{-5}	0.027
Ozone depletion	kg CFC-11 eq.	0	5.30×10^{-12}	41.7×10^{-12}	738×10^{-12}
Human toxicity potential	kg 1,4-DB eq.	0	1.8	30.6	63
Photochemical oxidant formation	kg NMVOC	0.0431	0.353	2.7	4.39
Freshwater ecotoxicity	kg 1,4-DB eq.	1.25×10^{-11}	1.26	0.0395	2.51
Fossil depletion	kg oil eq.	1.4	67.6	550	701

From the results above, it seems that the environmental performance of ramie fiber is not so competitive compared with glass fiber, especially when the spinning process was included into the whole manufacturing system of the ramie fiber. However, this idea may not be true. Because, at first, compared with the manufacturing technology of glass fiber, the production techniques of the ramie fiber are still not perfect and there is still a lot of room for improvement in the production techniques of ramie fiber. Then, there are no production techniques just for RFRP. For example, during the degumming process of the ramie fiber, when ramie fibers are used in FRPs, there may be no need for bleaching, which will reduce the consumption of sodium hypochlorite, washing water, and the releasing of treated wastewater. Then, the residual gum rate of the degummed ramie fiber should be less than 2% in the textile industry, while in the FRPs, the residual gum rate may be not so strict, which means that less chemicals may be used in the degumming process. Next, the spinning requirements for the textiles and FRPs may be different too. Finally, ramie fibers can degrade automatically, which can reduce the energy required for handling the disposals made of RFRP.

Based on the discussion above, more research should be performed to establish the manufacturing technologies of the ramie fiber just for RFRP. Degumming and spinning processes are the main processes that can be targeted to decrease the environmental burdens of ramie fibers used for FRPs.

Although some results have been obtained, there are also some limitations in this article. Firstly, the data used in the calculation is relatively limited, which means there will be more uncertainty regarding the final results. Then, data from the cooperative ramie textile mill was collected by the local workers in the mill, because of commercial confidentiality, which means there may be some variation because they know nothing about LCA principles. However, this study did provide some useful information and conclusions in general.

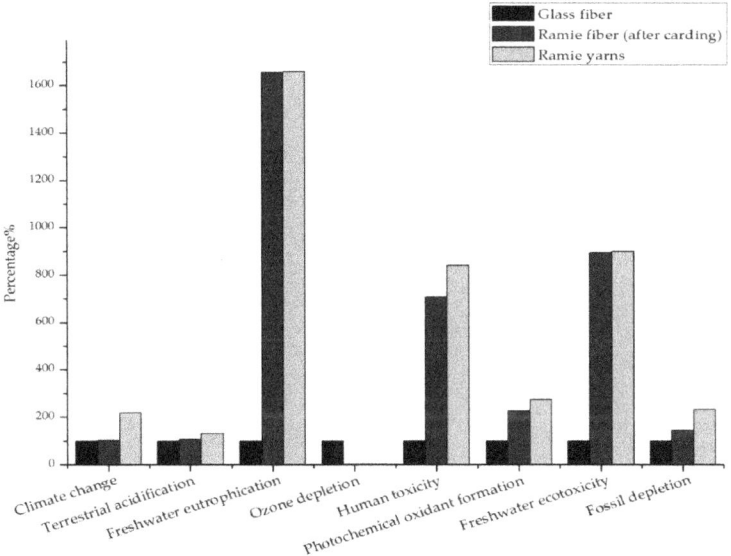

Figure 3. Comparison between three different kinds of fibers in terms of the eight main environmental categories.

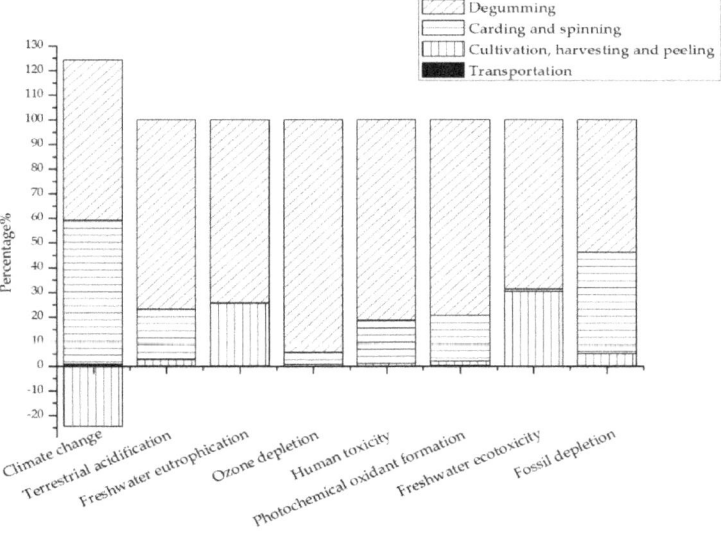

Figure 4. Percentages of different processes that contribute to the eight different environmental impact categories of ramie yarns.

4. Conclusions

According to LCA, the input and output data of the production processes of ramie fibers were summarized and analyzed. The comparison of LCA between the ramie fibers and glass fibers were performed. The following conclusions can be drawn based on the above study.

According to the results of LCA, if the spinning process was included within the manufacturing of the ramie fiber, the ramie fiber performs better than glass fiber only in ozone depletion. If the spinning process was not included, the ramie fiber performs better in the ozone depletion and has almost the same values of climate change and terrestrial acidification.

The degumming, and carding and spinning processes are the main causes of pollution during the production of ramie fibers. The cultivation, harvesting, and peeling processes mainly contribute to the categories of freshwater eutrophication, ozone depletion, and freshwater ecotoxicity. The transportation process contributes little to the overall pollution.

More research should be done toward developing production techniques for the ramie fibers for the RFRP.

Author Contributions: G.X. and X.-S.Y. conceived and guided the study; S.D. collected the data and wrote the paper.

Funding: This work was financially supported by the Chinese MIIT Special Research Plan on Civil Aircraft, grant No. MJ-2015-H-G-103 and European Union's Horizon 2020 research and innovation programme, under grant agreement No. 690638.

Conflicts of Interest: The authors declare no conflict of interest.

References

1. Mallick, P.K. *Fiber-Reinforced Composites: Materials, Manufacturing, and Design*, 3rd ed.; CRC Press: Boca Raton, FL, USA; London, UK; New York, NY, USA, 2007; ISBN 9780849342059-CAT# 4205.
2. Wötzell, K.; Wirth, R.; Flake, M. Life cycle studies on hemp fibre reinforced components and ABS for automotive part. *Die Angew. Makromol. Chem.* **1999**, *272*, 121–127. [CrossRef]
3. Schmehl, M.; Müssig, J.; Schönfeld, U.; Von Buttlar, H.B. Life Cycle Assessment on a Bus Body Component Based on Hemp Fiber and PTP. *J. Polym. Environ.* **2008**, *16*, 51–60. [CrossRef]
4. Scarponi, C.; Messano, M. Comparative evaluation between E-Glass and hemp fiber composites application in rotorcraft interiors. *Compos. Part B Eng.* **2015**, *69*, 542–549. [CrossRef]
5. Foraboschi, P. Analytical model to predict the lifetime of concrete members externally reinforced with FRP. *Theor. Appl. Fract. Mech.* **2015**, *75*, 137–145. [CrossRef]
6. Foraboschi, P. Effectiveness of novel methods to increase the FRP-masonry bond capacity. *Compos. Part B Eng.* **2016**, *107*, 214–232. [CrossRef]
7. Le Duigou, A.; Davies, P.; Baley, C. Environmental impact analysis of the production of flax fibres to be used as composite material reinforcement. *J. Biobased Mater. Bioenergy* **2011**, *5*, 153–165. [CrossRef]
8. Summerscales, J.; Dissanayake, N.P.J. Allocation in the life cycle assessment (LCA) of flax fibres for the reinforcement of composites. In Proceedings of the 21st International Conference on Composite Materials, Xi'an, China, 20–25 August 2017.
9. Islam, M.S.; Ahmed, S.K. The impacts of jute on environment: An analytical review of Bangladesh. *J. Environ. Earth Sci.* **2012**, *2*, 24–32.
10. Broeren, M.L.M.; Dellaert, S.N.C.; Cok, B.; Patel, M.K.; Worrell, E.; Shen, L. Life cycle assessment of sisal fibre—Exploring how local practices can influence environmental performance. *J. Clean. Prod.* **2017**, *149*, 818–827. [CrossRef]
11. Saheb, D.N.; Jog, J.P. Natural fiber polymer composites: A review. *Adv. Polym. Technol.* **1999**, *18*, 351–363. [CrossRef]
12. Liu, F.; Liu, Q.; Liang, X.; Huang, H.; Zhang, S. Morphological, anatomical, and physiological assessment of ramie [Boehmeria nivea (L.) Gaud.] tolerance to soil drought. *Genet. Resour. Crop. Evol.* **2005**, *52*, 497–506. [CrossRef]
13. Bo, W.; Dingxia, P.; Lijun, L.; Zhenxia, S.; Na, Z.; Shimei, G. An effcient adventitious shoot regeneration system for ramie (Bohehmeria nivea Gaud) using thidiazuron. *Bot. Study* **2007**, *48*, 173–180.
14. Thinkstep GaBi. Available online: www.gabi-software.com (accessed on 24 April 2018).
15. ISO 14040:2006 Environmental Management—Life Cycle Assessment—Principles and Framework. Available online: www.iso.org (accessed on 24 April 2018).

16. ISO 14044:2006 Environmental management—Life cycle Assessment—Requirements and Guidelines. Available online: www.iso.org (accessed on 24 April 2018).
17. Paul, W.; Ivens, J.; Ignaas, V. Natural fibres: Can they replace glass in fibre reinforced plastics? *Compos. Sci. Technol.* **2003**, *63*, 1259–1264.
18. Ekvall, T.; Finnveden, G. Allocation in ISO 14041—A critical review. *J. Clean. Prod.* **2001**, *9*, 197–208. [CrossRef]
19. Ramie Institute of Hunan Agricultural University. Available online: http://chinagrass.hunau.edu.cn/ (accessed on 24 April 2018).
20. Wang, C.T.; Li, Z.D.; Yu, T.W.; Liu, M.S.; Xiao, Z.P.; Cui, G.X.; Xue, N.D.; Li, G.S.; Deng, M.Q. Fertilization programma in ducing fine quality and high yield of ramie and its application technique. *J. Hunan Agric. Coll.* **1994**, *20*, 318–324.
21. Dissanayake, N.P.J. Life Cycle Assessment of Flax Fibres for the Reinforcement of Polymer Matrix Composites. Ph.D. Thesis, University of Plymouth, Plymouth, UK, May 2011.
22. Ministry of Agriculture of the People's Republic of China. Available online: www.moa.gov.cn (accessed on 24 April 2018).
23. 2006 IPCC Guidelines for National Greenhouse Gas Inventories—Volume 4—Agriculture, Forestry and Other Land Use. Available online: www.ipcc-nggip.iges.or.jp/public/2006gl/index.html (accessed on 24 April 2018).
24. EU Concerted Action Report Harmonization of Environmental Life Cycle Assessment for Agriculture Final Report. Concerted Action AIR3-CT94-2028. Available online: www.researchgate.net (accessed on 24 April 2018).
25. Bioenergy for Europe: Which One Fits Best? A Comparative Analysis for the Community. Available online: https://www.researchgate.net/publication/320258744_Bioenergy_for_Europe_Which_Ones_Fit_Best_-_A_Comparative_Analysis_for_the_Community (accessed on 24 April 2018).
26. Huang, J.C.; Li, X.W.; Zhang, B.; Tian, K.P.; Shen, C.; Wang, J.G. Research on the 4LMZ160 crawler ramie combine harvester. *J. Agric. Mech. Res.* **2015**, *9*, 155–158. [CrossRef]
27. Changchai Company Limited's Product Sheet. Available online: http://www.changchai.com.cn/show.asp?id=18 (accessed on 24 April 2018).
28. Benoît, C.; Xavier, P.; André, G.; Fritz, S.; Gérard, C. Contributions of hemicellulose, cellulose and lignin to the mass and the porous properties of chars and steam activated carbons from various lignocellulosic precursors. *Bioresour. Technol.* **2009**, *100*, 292–298.
29. Shengwen, D.; Zhengchu, L.; Xiangyuan, F.; Ke, Z.; Lifeng, C.; Xia, Z. Diversity and characterization of ramie-degumming strains. *Sci. Agricola* **2011**, *69*, 119–125.
30. Deb, P.R.; Pratik, S.; Sabyasachi, M.; Pradipta, B.; Rakesh, K.G. Degumming of ramie: Challenge to the queen of fibres. *Int. J. Bioresour. Sci.* **2014**, *1*, 37–41.
31. Bao, H.F.; Bao, J.G.; Li, H.W. Treatment of ramie retting wastewater using acidifying hydrolysis/biological oxidation/photochemical process. *China Water Wastewater* **2006**, *22*, 52–55.

Article

Effect of Paper or Silver Nanowires-Loaded Paper Interleaves on the Electrical Conductivity and Interlaminar Fracture Toughness of Composites

Miaocai Guo [1] and Xiaosu Yi [1,2,*]

[1]   National Key Laboratory of Advanced Composites, AVIC Composite Technology Center, Beijing 101300, China; guo_miaocai@sina.cn
[2]   Faculty of Science & Engineering, University of Nottingham Ningbo China (UNNC), Ningbo 315100, China
*   Correspondence: yi_xiaosu@sina.cn or xiaosu.yi@nottingham.edu.cn; Tel.: +86-574-8818-8746

Received: 19 June 2018; Accepted: 17 July 2018; Published: 19 July 2018

Abstract: The effect of plant-fiber paper or silver nanowires-loaded paper interleaves on the electrical conductivity and interlaminar fracture toughness of composites was studied. Highly conductive paper was prepared by surface-loaded silver nanowires. The percolation threshold appeared at about 0.4 g/m^2. The surface resistivity reached $2.3 \ \Omega/\text{sq}$ when the areal density of silver nanowires was 0.95 g/m^2. After interleaving the conductive papers in the composite interlayers, in-plane electrical conductivity perpendicular to the fiber direction was increased by 171 times and conductivity through thickness direction was increased by 2.81 times. However, Mode I and Mode II interlaminar fracture toughness decreased by 67.3% and 66.9%, respectively. Microscopic analysis showed that the improvement of conductivity was attributable to the formation of an electrical conducting network of silver nanowires which played a role in electrical connection of carbon fiber plies and the interleaving layers. However, the density of the highly packed flat plant fibers impeded the infiltration of resin. The parallel distribution of flat fibers to the carbon plies, and poor resin-fiber interface made the interlaminar fracture occur mainly at the interface of plant fibers and resin inside the interleaves, resulting in a decline of the interlaminar fracture toughness. The surface-loading of silver nanowires further impeded the infiltration of resin in the densely packed plant fibers, resulting in further decline of the fracture toughness.

Keywords: functional composites; electrical properties; fracture toughness; function integrated interleave; plant fiber; paper

---

1. Introduction

Continuous carbon fiber-reinforced resin matrix composites are usually made of continuous carbon fibers as the reinforcement and thermosetting resin as the matrix. Specifically because of their high strength and modulus, these materials are widely used in aerospace and attract an increasing amount of attention in many fields [1]. On the other hand, going green and going functional are the two major themes in material development in the world. To develop greener composites, an increasing number of researchers are focused on the 'greening' of raw materials such as developing bio-based epoxy resin matrix and applying natural fibers as the reinforcement [2–4]. To develop functional composites while maintaining their performance, many studies focus on the application of nanomaterials, for instance by improving the conductivity of composites for lightning-strike protection and electromagnetic shielding [5–7].

One of the most effective ways to improve the impact resistance of composites is through interleaving technology. Currently, interleaving materials generally use thermoplastic resin particles, thin films and non-woven fabrics [8–11]. Among these materials, tough and high-strength fibers such as nylon fiber, aramid fiber, carbon nanofiber and their non-wovens are attracting more attention [9,12–14]. By introducing micro- and nano-structures into the interlays, new crack propagation mechanisms were introduced to improve the areal density of energy dissipation, thus the interlaminar fracture toughness of the composites was remarkably improved. However, the structures of the interleaves strongly affect the interlaminar toughness of the composite. Palazzetti et al. studied the effects of thickness, orientation and diameter of electrospun nylon6,6 veils on the Mode I (G_{IC}) and Mode II (G_{IIC}) interlaminar fracture toughness of composites [15]. Ramirez et al. studied the interlaminar fracture toughness of composites interleaved with PPS and PEEK veils of different areal densities, linear densities and fiber diameters. They found that G_{IC} and G_{IIC} were increased with the mean coverage, but the polymer types showed nearly no effect [16]. Heijden et al. studied the toughening properties of polycaprolactone interleaves with different non-porous and porous structures. They found the polycaprolactone thin layer of nanofibers had the best performance [17]. KuWata et al. studied the toughening properties of five interleaf veils formed by carbon fibers, polyester fibers, and polyamide fibers and their mixtures [18,19]. Beckermann et al. found that the most important factors of the interleaf layers were the polymer types and areal densities [20]. Daelemans et al. found the orientation of veil fibers strongly affected toughening performance of veils. The laminate interleaved with veils with random fiber orientation had the highest interlaminar fracture toughness [21]. All these studies revealed that the material property, fiber diameter, areal densities, fiber orientation and other structural factors of the interleaved layer showed large or small effect on the toughening performance related to the composite material systems. However, although many studies have been done, the interlayer toughening mechanism is still not thoroughly understood [22].

Combined with the interleaving technology, a functionalized interlayer technology was developed in our previous researches which can simultaneously and effectively improve the interlaminar fracture toughness and the electrical conductivity of composites [23,24]. The principle was constructing an electrical conducting–toughening double functions network in the interlays of composites by interleaving thin sheet formed of micron-sized nylon network and nano-sized silver nanowires network. The loaded highly conductive silver nanowires showed no negative influences on the toughening mechanism, thus the toughness performance of the interleave was well kept. Aside from in our work, the nano-hybrid of interleaf has attracted much attention elsewhere in recent years. Eskizeybek et al. used electrospinning carbon nanotube-polyacrylonitrile (CNT-PAN) hybrid nanofibrous mat as the interleaf to toughening composite. The G_{IC} of the composite was much higher than that interleaved with neat PAN nanofibrous mat [25]. Zhou et al. used hierarchical CNT-short carbon fiber interleaves to improve the G_{IC} of composite [26]. Zheng et al. used carbon nanotubes/polysulfone nanofiber prepared by vacuum filtration as the toughening interleaf [27]. Lee et al. used CNT-enhanced nonwoven carbon tissue as the interleaf and found a significant improvement of G_{IIC} [28]. Although the functions of the materials in these references were not studied, the multi-scale toughening mechanism of different material systems was interesting and still needed to be further explored.

The above also indicates most current research is focused on tough and high-strength artificial fibers. However, although many plant fibers have good mechanical properties and have been used as reinforced fibers [29], the use of plant fibers or papers as interleaving materials to modify composites is rarely reported. Moreover, different from the circular shaped artificial fibers reported before, plant fibers have extremely complex structures, with most of them containing two walls and lumens [30]. Study of the effect of different structures and materials of plant-fiber interleaves can give us a deeper understanding of the interlaminar failure mechanism of composites. The nano-scale effect on the toughening mechanism and the function integration of the composite using nano hybridized plant-fiber interleaf are also interesting.

In this paper, a conductive plant fiber paper was prepared through solution immersion method. Using the papers and conductive papers as the interleaving materials, the interlaminar fracture toughness and conductivity of the modified composites was studied, and the influence mechanism of the multi-scale structures on the interlaminar properties is discussed.

2. Materials and Methods

2.1. Materials

In this study, papers used as interleaves were a kind of thin layer formed by plant fibers, with the main component being cellulose. The thickness of the paper was 16~18 µm. The surface density was 11.26 g/m^2. Silver nanowires (AgNWs) were purchased from Beijing NaHui Technology and Trade Co., Ltd. (Beijing, China). The diameter of the silver nanowire was about 40 nm. The length was between 30 µm and 50 µm. The silver nanowires were dispersed in isopropyl alcohol in a concentration of 5 mg/mL. The unidirectional carbon fiber fabric used in this study was U3160 (CCF 300, 3K, areal density: 160 g/m^2). The thickness of a single ply was 0.166 mm. The average diameter of the carbon fiber was 7.2 ± 0.3 µm. An aero-grade RTM (resin transfer molding) epoxy resin under the brand 3226, which is a product of AVIC Composites Technology Co., Ltd. (Beijing, China) was chosen as the matrix resin. All other auxiliary materials were purchased from commercial sources.

2.2. Preparation of the Conductive Papers and the Composite Laminates

Figure 1 illustrates the preparation process of the conductive papers, composite laminates, and test samples. First, the conductive paper was prepared by immersion method. Then the unidirectional carbon plies were stacked into preforms with one sheet of plain or conductive paper inserted as the interleaf in each interlayer. Finally, the required composites were injected with epoxy resin 3226 and cured at the given RTM molding condition of U3160/3226 composite. The test specimens were prepared according to the corresponding test standards.

To prepare the conductive papers, the plain papers were immersed into the slurry of AgNWs dispersed by isopropanol. After immersion for 5 s at room temperature, the papers were removed from the slurry and dried at room temperature. The conductive papers of different AgNWs areal densities can be obtained by controlling the immersion times. The conductive paper used in this study had an AgNWs areal density of 0.95 g/m^2 obtained by performing the immersing and drying process twice.

The composite laminates prepared for the conductivity test and interlaminar fracture toughness test were unidirectional with the preform stacked sequence of $[0]_{24}$ according to specimens for testing (Figure 2). Here $[0]_{24}$ means total 24 carbon fiber plies were stacked with all the fibers oriented at 0° direction. Each interlayer of the preform was interleaved with one plain paper or conductive paper. To prepare the samples for G_{IC} and G_{IC} tests, a 25 µm thick polytetrafluoroethylene (PTFE) film was inserted into the middle plane of the preform to prefabricate the cracks with controlled length. After being prepared, the preforms were injected with 3226 epoxy resin and cured according to the standard RTM method of U3160/3226 composite. After cooling, the composite laminates were released from the mold, and then the samples were machined according to the test standards.

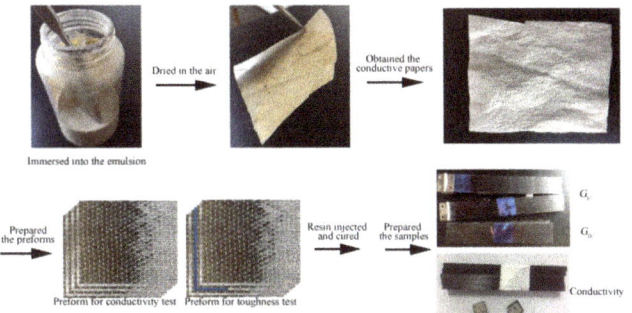

Figure 1. Schematic diagram of preparation of conductive papers, composite laminates and test specimens.

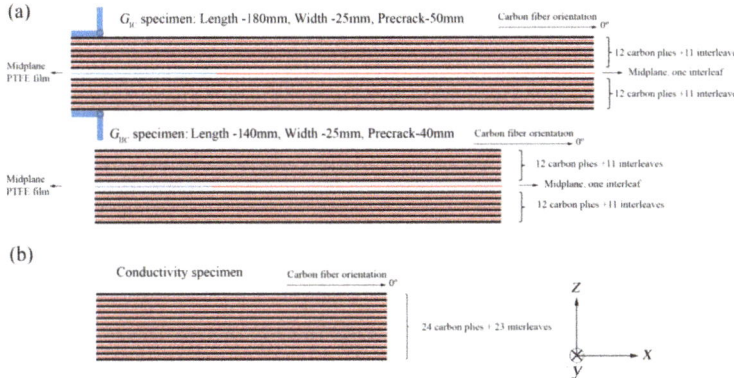

Figure 2. Schematic map of the specimen geometries for fracture toughness and conductivity testing. (a) G_{IC} and G_{IIC} specimens; (b) conductivity specimen.

2.3. Measurement of Mode I and II Fracture Toughness

The characterization method of fracture toughness and conductivity was the same as our previous paper [23]. The specimen geometry is given in Figure 2a. Mode I fracture toughness tests were carried out according to the Chinese Aviation Industry Standard HB 7402 96 using double cantilever beam (DCB) specimens. The specimens had a width of 25 mm and a length of 180 mm. A pre-crack with a length of 50 mm was made by inserting a PTFE film in the midplane. To make a fresh pre-crack, an initial loading was applied to the specimen before testing. The growing length of the fresh pre-crack was controlled to around 20 mm. The specimen was then reloaded, and the loading was stopped after an increment of a delamination crack growth of 10 mm. The process was carried out five times. Three specimens were tested for each sample. G_{IC} was calculated from Equation (1).

$$G_{IC} = \frac{mP\delta}{2Wa} \times 10^3 \qquad (1)$$

where:

G_{IC} = Mode I interlaminar fracture toughness, J/m^2
m = coefficient
P = load, N
δ = load point displacement, mm

W = specimen width, mm

a = delamination length, mm.

Mode II fracture toughness tests were performed according to the Chinese Aviation Industry Standard HB 7403-96. The specimens had a width of 25 mm and a length of 180 mm. The length of pre-crack made by a PTFE film was 40 mm. To make a fresh pre-crack, an initial loading was applied to the specimen before testing first. The growth of delamination crack was about 5 mm. The specimen was then also reloaded. Five specimens were tested. G_{IIC} was calculated from Equation (2). A difference is that the crack growth of the G_{IC} test was stable, while the growth of the G_{IIC} test was unstable and faster propagated.

$$G_{\text{IIC}} = \frac{9P\delta a^2}{2W(2L^3 + 3a^3)} \times 10^3 \tag{2}$$

where:

G_{IIC} = Mode II interlaminar fracture toughness, J/m^2

P = load, N

δ = load point displacement, mm

a = delamination length, mm

W = specimen width, mm

$2L$ = support span, mm.

2.4. Measurement of Conductivity

The AgNWs-loaded papers used for the surface resistivity test were 100 mm wide and 100 mm long. The two opposite edges of the paper were silver-painted and then dried. When testing, copper plates were tightly pressed on the silver-painted edges as electrodes. The surface resistivity (R_s) was calculated from the test resistance (R) by Equation (3). The volume resistivity (R_v) was calculated by Equation (4).

$$R_s = R \times \text{width}/\text{length} \tag{3}$$

$$R_v = R \times \text{width} \times \text{thickness}/\text{length} \tag{4}$$

Three main directions of the composite for the conductivity test were defined in Figure 2b since most fiber materials had their orientation [31]. The in-plain volume conductivity along the fiber direction (x direction) and perpendicular to the fiber direction (y direction) of the carbon composite laminate was tested on the samples 10 mm wide and 100 mm long. The volume conductivity through-thickness direction (z direction) was measured on the samples 5 mm square. For each orientation five samples were tested. The two opposite surfaces of the composite specimens were silver-painted and dried before test. When testing, two copper plates were tightly pressed on the silver-painted surface as electrodes. The volume resistivity and conductivity (σ) was calculated by Equations (4) and (5) respectively.

$$\sigma = 1/R_v \tag{5}$$

2.5. Characterization of Morphology

Scanning electron microscopy (SEM) images were obtained using a Hitachi S-4800 SEM (Hitachi, Ltd., Tokyo, Japan). All samples for SEM test were coated with a gold layer before the test. The photographs were taken using a smart phone. The optical micrographs were obtained by an Olympus SZ61 optical microscope (Olympus Corporation, Tokyo, Japan).

3. Results

3.1. Preparation of the Conductive Paper

Figure 3 gives the SEM images of the paper. The paper had a thickness of 16~18 μm and areal density of 11.26 g/m². The paper was formed from a large number of plant fibers overlapping each other. The plant fibers were flat-like and their widths were between 20 μm and 40 μm. A large number of throughout-thickness holes uncovered with fibers existed in the paper because areal density of the paper was small. Figure 3c,d give the cross section of the plant fibers. The plant fiber was hollow with a single wall. The wall thickness was about 1.5 μm. Due to the paper production process involved the extrusion process, all the plant fibers were flattened. Two typical shapes of the flattened fibers are given in Figure 3c,d. The fiber thickness was about 3~5 μm. Figure 3e gives the cross section of the plant fibers embedded in the epoxy resin. It also shows that the thickness of the fiber wall was about 1.5 μm. Some of the plant fibers which were not completely flattened had a larger apparent thickness and epoxy resin was filled into the hollow cores. For those plant fibers completely flattened, the two walls were tightly contacting each other. As seen from Figure 3e, the flattened fibers overlapped each other and the resin had difficulty penetrating their gaps. In addition, the interface bonding between fiber and resin was obviously separated, indicating that interface adhesion was poor.

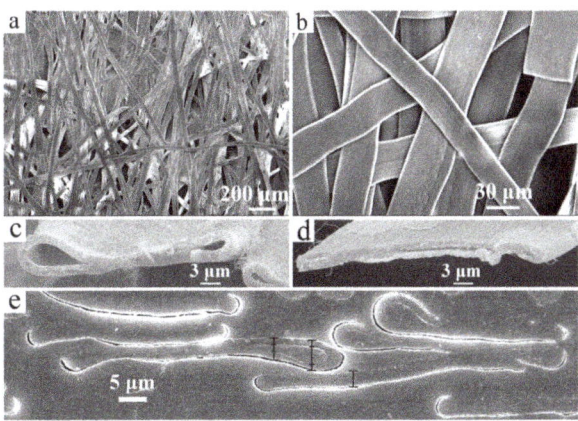

Figure 3. SEM images of (**a,b**) paper; (**c,d**) cross section of a plant fiber; (**e**) cross section of paper embedded in epoxy resin.

The conductive paper was obtained by surface-loading AgNWs on the paper using an immersion method which is widely used to prepare functional thin sheets or porous materials such as conductive films, non-woven fabrics, and foams etc. [24,32,33]. Generally, the thin sheet was dipped into the slurry of nanomaterials to absorb a certain amount of dispersive liquid. After being dried, the surface of the thin sheet was covered with a layer of nanomaterials and thus it was functionalized. After being loaded with AgNWs, the color of the paper changed from white to silver gray. However, the AgNWs were only adsorbed on the fiber surface and could be easily removed by ultrasonic treatments. Figure 4a gives the areal density of AgNWs changing with the dipped time. It indicates the areal density was linearly increased with the increase of immersion time. Figure 4b gives the surface resistivity of the paper changed with the areal density of AgNWs. Before being loaded with AgNWs, the paper was obviously electrical insulation material. However, after being loaded with AgNWs, the paper became conductive and its conductivity was gradually increased with the increase of AgNWs areal density. The percolation threshold was found to be near 0.3 g/m² areal density of AgNWs. To the paper loaded with 0.95 g/cm² AgNWs, the surface resistivity was typically 2.3 Ω/sq., that is, the volume resistivity was calculated to be 256 S/cm, indicating that the paper had good conductivity with a small AgNWs loading. The conductive paper was still very flexible, and met the requirements of composite processing.

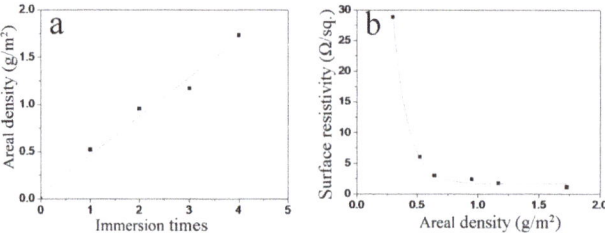

Figure 4. (a) Areal density of AgNWs versus immersion times in AgNWs slurry of paper; (b) Surface resistivity of paper versus areal density of AgNWs.

Figure 5 shows the SEM images of the paper after being loaded with AgNWs of different areal densities. Figure 5a gives a large scaled view of the AgNWs-loaded paper with the areal density of 0.95 g/m². It can be seen that AgNWs were uniformly adsorbed on the surface of the plant fibers, and almost covered all the fiber surface forming a continuous conducting network, thus the paper loaded with AgNWs has a low percolation threshold and good conductivity. Figure 5b–d give the SEM images of the conductive papers with different AgNWs areal densities. For all densities, AgNWs were uniformly distributed on the surface of the paper. With an increase of AgNWs density, the AgNWs network on the paper gradually went from sparse to dense. The plant fibers were barely observed at the AgNWs density of 1.73 g/m² (Figure 5d). However, no local enrichment of AgNWs was found. It also indicated AgNWs were uniformly distributed even at such a high loading. In addition, in the case of large holes that not covered with plant fibers on the paper (generally larger than 10 μm in length, seen in Figure 5a,c,d), AgNWs bridged across the holes under the capillary of the dispersion. The existence of AgNWs bridging made the AgNWs network distributed on the two opposite surfaces of the paper interconnect and formed a holistic continuous conducting structure. The above results show that we obtained the conductive paper with a continuous and uniform AgNWs conducting network on surface the paper.

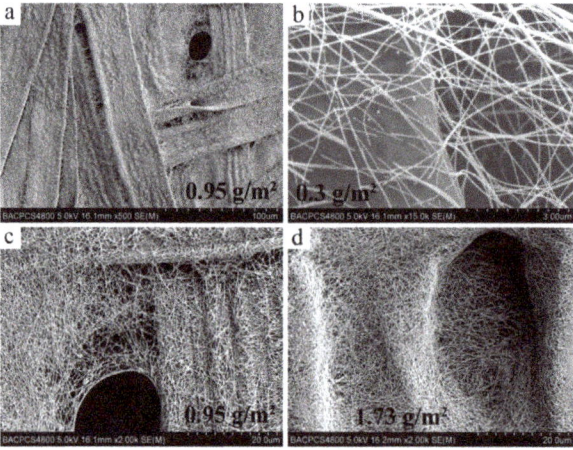

Figure 5. SEM images of the conductive papers with different AgNWs areal densities. (**a**) 0.95 g/m², a large scaled view; (**b**) 0.3 g/m²; (**c**) 0.95 g/m², a small scaled view; (**d**) 1.73 g/m².

3.2. Electrical Conductivity of the Laminates

Three groups of samples containing control, plain papers interleaved, conductive papers interleaved for conductivity test were prepared using RTM method. The stacking sequence of all the samples was $[0]_{24}$. The final thickness was controlled to 4.0 mm by steel spacers. Table 1 contains a summary of the volume electrical conductivity in the two in-plane directions (σ_x and σ_y) and the through-thickness direction (σ_z) of all these samples.

Table 1. Electrical conductivity of composite laminates samples studied.

Samples	In-Plain		Out of Plain	Fiber Volume Fraction V_f (%)
	σ_x (S/m)	σ_y (S/m)	σ_z (S/m)	
Control, no interleaf	~6×10^3	9.7	4.7	54.7
Interleaved with plain papers	~6×10^3	9.2	<1.3×10^{-6}	54.7
Interleaved with AgNWs-loaded papers	~6×10^3	1.67×10^3	17.9	54.7

The in-plane electrical conductivity along the fiber direction (σ_x) for all the composite samples was about 6×10^3 S/m. This indicated that the change of σ_x was very small and σ_x was mainly depended on the intrinsic conductivity of the carbon fibers because it was much higher than the conductivity of the interleaves.

The in-plane electrical conductivity perpendicular to the fiber direction (σ_y) of the conductive papers interleaved samples was significantly increased by 171 times and 180 times compared with the control and plain paper interleaved samples respectively. This means σ_y of the control and plain paper interleaved samples were very close, while σ_y of conductive papers interleaved sample was significantly higher.

The through-thickness conductivity of the conductive paper interleaved samples reached 17.9 S/m and also significantly increased 2.81 times compared with the control samples, while for the plain paper interleaved samples, the through-thickness direction was nearly insulated. The interleaves played barrier roles in the electrical conduction between carbon fiber layers, leading to the decrease of conductivity. When interleaved with the conductive papers, the laminated carbon fiber layers and the conductive papers formed a series structure. The through-thickness conductivity was mainly limited by the conductivity of the carbon fiber layers because the AgNWs-loaded conductive papers had relatively higher conductivity. The good conductivity of the conductive papers played an effective conducting connection, making the through-thickness conductivity of composite samples greatly improved.

Compared with our previous work [23], the in-plane and out of plain conductivities of composites were smaller. That is because the conductivity was influenced by many factors such as the structural characteristics of the conductive interleaves, the bulk conductivity of carbon fibers, the volume fraction of composites and so on.

Figure 6a,b give the cross section SEM images of the composites interleaved with plain papers and conductive papers. The interleaves were distributed between two carbon layers with thickness between 5 μm and 25 μm. The different contrast shown in Figure 6b can be assigned to the different phases/structures based on their electrical properties, e.g., non-conductive epoxy resin region was dark while the AgNWs region was much brighter [34,35]. Comparing Figure 6a,b, relatively bright zones on the surface of the plant fibers of the conductive paper were found due to the high conductivity of AgNWs in the epoxy resin (see the enlarged images of the regions 1 and 2). The loading of AgNWs on the surface of the paper showed nearly no effect on the thickness of the interlayer. The conducting paths on the opposite faces of the paper formed by AgNWs-doped epoxy resin were connected in the hole not covered with plant fibers shown in the enlarged image of Region 2. Thus, a continuous conducting structure on the whole surface of the interleave was formed. As seen from the enlarged image of Region 1, the conducting path had good electrical contact with carbon fibers, so it played a role in connecting the two carbon layers. AgNWs were well and uniformly distributed on the surface of paper, indicating the conducting nano-sized network was well kept after stacking, resin injection and curing process. Figure 6c is the cross section image of the composite with an open crack. After wearing during the sample preparation process, the plant fiber was stripped from the matrix resin, indicating a poor bonding between the plant fiber and the resin which was agreed with the poor interface shown in Figures 3e and 6a,b . Usually this was because the plant fibers were hydrophilic and incompatible with epoxy resin.

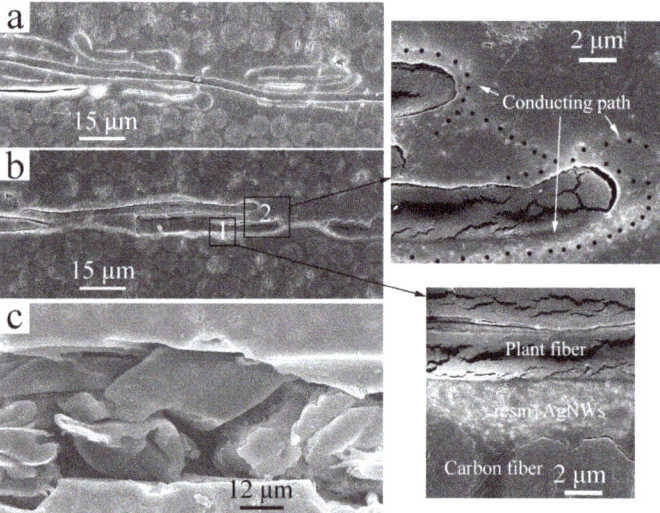

Figure 6. SEM images of the cross section of the interleaved composites. (**a**) Interleaved with plain paper; (**b**) Interleaved with conductive paper. Region 1: An enlarged image of the area between carbon fibers and plant fiber. Region 2: The conducting path formed by AgNWs; (**c**) Polished plain paper interleaved sample with an open crack.

3.3. Interlaminar Fracture Toughness of the Laminates

Table 2 shows the interlaminar fracture toughness of all the samples in terms of Mode I (G_{IC}) and Mode II (G_{IIC}), respectively. The final thicknesses of all laminates were controlled to be around 4.0 mm, which satisfies the nominal carbon fiber volume fraction. However, after being interleaved with plain papers or conductive papers, Mode I and Mode II interlaminar fracture toughness decreased significantly. G_{IC} decreased from 321.1 J/cm^2 to 111.1 J/cm^2 (−65.4%) and 104.8 J/cm^2 (−67.4%) and G_{IC} decreased from 1293 J/cm^2 to 541 J/cm^2 (−58.2%) and 428 J/cm^2 (−66.9%) respectively. This indicated that the interleaves did not toughen the composites; however, they played a "defect" role in the interlayers. The interlaminar fracture toughness was further smaller to the AgNWs-loaded papers interleaved composite.

Table 2. Mode I and II fracture toughness of non-interleaved and interleaved laminates.

Samples	G_{IC} (J/m^2)	G_{IIC} (J/m^2)
Control, no interleaf [1]	321.1	1293
Interleaved with plain papers	111.1 ± 2.6	541 ± 55
Interleaved with AgNWs-loaded papers	104.8 ± 6.7	428 ± 19

[1] Data from the internal report of our lab.

Figure 7 gives the optical images and micrographs of the Mode I and Mode II fracture surfaces of the papers and conductive papers interleaved composites. As can be clearly seen from the figure, all the fracture surfaces occurred inside the interleaves, which means a similar fracture mechanism for the different interleaves. In addition, the fracture surface of the conductive paper interleaved composites was relatively whitish, probably because of the existing of AgNWs or the insufficient infiltration of epoxy inside the conductive paper.

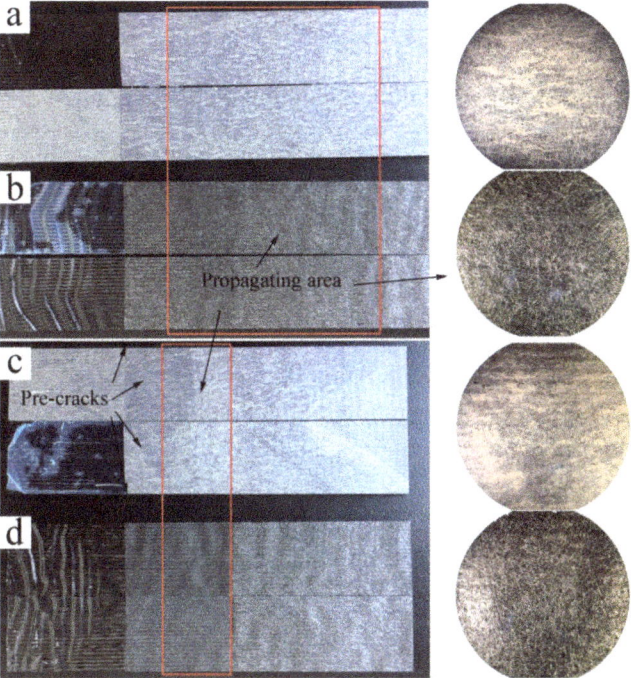

Figure 7. Optical photos and optical microscope photographs of failure surfaces after Mode I and Mode II fracture toughness test. (**a**) conductive papers interleaved, Mode I fracture surface; (**b**) plain papers interleaved, Mode I fracture surface; (**c**) conductive papers interleaved, Mode II fracture surface; (**d**) plain papers interleaved, Mode II fracture surface.

Figure 8 gives the SEM images of the Mode I fracture surfaces. It was found that all the Mode I fracture surfaces were covered with plant fibers, that is, the failure mainly occurred inside the interleave layers. Compared with the structure of nylon veil [23], the flattened plant fibers of the paper stacked more densely. More fibers were distributed per unit volume, and more parallel to the carbon plies. From Figure 8a,c, most of the interlaminar failures occurred in the plant fiber–epoxy resin interfaces or the overlapped areas of plant fibers. Compared with Figures 3e and 6a,b , because of the flat shape and highly packed density of the plant fibers, we can see the epoxy resin had not completely penetrated into the overlapped areas of the plant fibers. In addition, poor interface between plant fiber and resin was found, thus it was much easier for the fracture to occur inside the interleaves. Besides, the surface of plant fibers or resin on the failure surface was very smooth, and no destroyed resin particles were found, also indicating that the interfacial bonding between paper fiber and resin was weak. Only a few plant fibers were destroyed, as can be seen in Figure 8b,d. In the case of the AgNWs-loaded paper interleaved sample, AgNWs can easily be seen at the resin area of fracture surface shown in Figure 8d. However, this area has became more rough indicating that the resin had relatively poor infiltration in the AgNWs-enriched area.

Compared with Figure 8a,c, resin may be easily seen in the gap of the overlapped plant fibers in Figure 8a, while the resin was small in quantity except for the area not covered with plant fiber, and more bare plant fibers are found in Figure 8c. This indicates that the presence of AgNWs concentrated on the surface of the interleaved paper greatly reduced the permeability of the resin into the densely packed plant fibers, making the interlayer much liable to be destroyed when not filled with resin, and thus the Mode I fracture toughness of the conductive papers interleaved composite was lower than the plain papers interleaved composite.

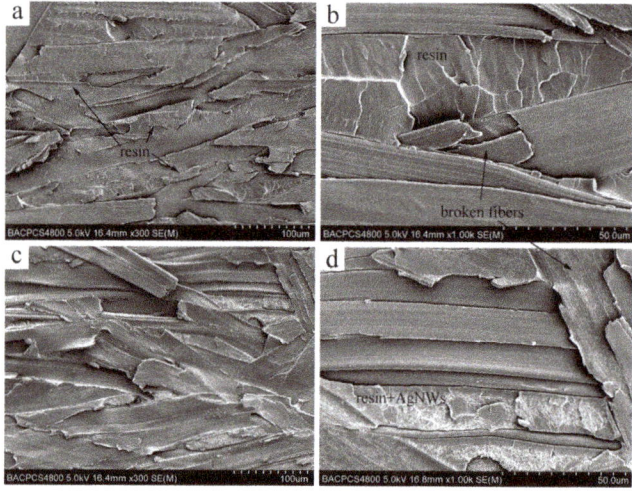

Figure 8. SEM images of Mode I fracture surfaces of (**a**,**b**) plain paper interleaved composite and (**c**,**d**) conductive paper interleaved composite.

Figure 9 gives the SEM images of the Mode II fracture surfaces. It can also be seen from the images that most of the Mode II fractures occurred inside the interleave layers of the composites. Only in the area not covered with plant fibers, shear failure of resin was observed at the interface between the carbon ply and the interlayer. As seen from Figure 9a,c, fracture failure at plant fiber–resin interfaces, paper fiber–paper fiber interfaces obviously exist everywhere. As seen from Figure 9b,d, the fracture failure of plain paper interleaved composite mostly occurred at the fiber-resin interfaces while that of conductive paper interleaved composite mostly occurred at the non-resin infiltrated fiber–fiber overlapped areas. More surfaces of plant fibers were found and little resin existed in the gap of plant fibers as seen in Figure 9c,d, which are very different from the fracture surfaces shown in Figure 9a,b. This indicates that the presence of AgNWs impeded the impregnation of resin in densely packed paper fibers, leading to the further decline of Mode II interlaminar fracture toughness.

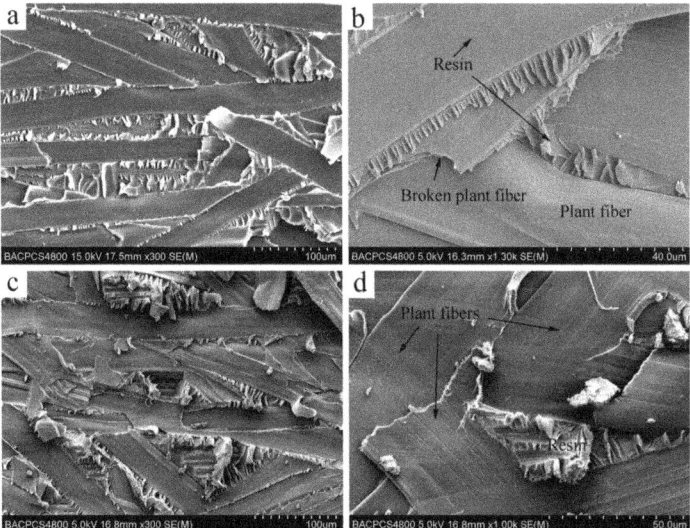

Figure 9. SEM images of Mode II fracture surfaces of (**a**,**b**) plain papers interleaved composite and (**c**,**d**) conductive papers interleaved composite.

4. Discussion

It was proved that the electrical conductivity and fracture toughness of the composite can be simultaneously improved using the conducting–toughening integrated interleaves [23]. However, from the results in this paper, the structure of function integrated interleaves still requires more study.

To improve the electrical conductivity, the continuity of the conducting network which played a role in electrical connection of different carbon fibers was very important. On the surface of the conductive paper in this study, AgNWs formed a continuous and uniform conducting network, making the paper highly conductive at a low AgNWs loading. The porous form of the paper made the conductive networks on the opposite surfaces of the paper interconnected, thus the interleaves effectively connected the different carbon plies, resulting in improvement of the through-thickness conductivity of the composite.

With the interlaminar toughening of composites, a fracture usually occurs in the path of least energy dissipation. The toughening performance of the interleave is closely related to its material properties and structural properties. The toughening mechanism generally includes viscoelastic energy dissipation, bridging the effect of particles or fibers, and crack deflection caused by defects or rigid interfaces. When a fracture path propagates through areas with these effects, energy dissipation per unit area is increased and the fracture path becomes more circuitous, resulting in an increase of fracture toughness. For example, as a successful application of ex situ toughening, the toughening mechanism of PEK-C film was the formation of a double continuous structure containing thermoplastic rich phase and thermosetting rich phase. The higher viscoelastic energy dissipation of thermoplastic rich phase and the crack deflection between two-phase interface increased energy dissipation of the crack propagation. With the nylon veil interleaved composites, the bridging effect of ductile nylon fibers was the main factor.

Compared with the circular artificial nylon fibers, the plant fibers were flattened. This made the packing density much higher and the plant fiber orientation was more parallel to the carbon plies. Epoxy resin had more difficulty penetrating into such closely packed narrow gaps between plant fibers. Less fiber bridging also occurred because of its parallel orientation during fracture. On the other hand, because of the strongly hydrophilic characteristic of plant fiber, the interface between plant fiber and epoxy resin was poor. Combined the above factors, the interlaminar failure mostly occurred in fiber-resin interface and non-resin infiltrated fiber–fiber overlapped area where the energy dissipation per unit area was much lower, thus the interlaminar fracture toughness was significantly decreased. With the AgNWs-loaded papers interleaved composites, a large number of AgNWs formed a nanoscaled network on the surface of densely packed plant fibers, which was very different from the AgNWs-loaded nylon veil interleaved composite, with which the AgNWs network mostly covered single nylon fibers and bridged between fibers, further reducing the penetration of resin in the interior of paper, so interlaminar failure occurred in non-resin infiltrated fiber–fiber overlapping areas and led to the further decline of interlaminar fracture toughness.

In addition, although interlaminar toughness was decreased after interleaving with papers made of plant fibers, the result can provide us with some useful inspirations. For example, we could use flax fiber, which is much stronger and nearly circular to improve the fiber fracture energy and infiltration of resin. Additionally, the surface treatment is required to improve the bonding of fiber and resin. The sound absorption and damping properties of composites interleaved with plant fibers may also be very interesting.

5. Conclusions

In conclusion, through preparing conductive papers and using as interleaves, the electrical conductivity of composites was effectively improved, but the interlaminar fracture toughness was obviously decreased. The increase of conductivity was achieved from the good conductivity of AgNWs network and good electrical connection between AgNWs network and carbon plies. The decrease of the interlaminar fracture toughness can be attributed to highly parallel, densely packed interleaf structure formed by the flattened plant fibers and the poor interface between plant fibers and epoxy resin. The highly parallel distribution of plant fiber and its poor bonding with resin caused the fracture path to mainly propagate on the peeling surfaces of epoxy resin–plant fiber and plant fiber–plant fiber. Few plant fibers were destroyed during facture to dissipate energy. The densely packed plant fibers impeded the penetration of resin inside the paper and the penetration became worse for the paper with AgNWs surface loaded. All these reasons resulted the decrease of interlaminar fracture toughness.

Author Contributions: X.Y. conceived and guided the project and study; M.G. performed the experiments and wrote the paper.

Funding: Financial supports from National 973 Project 2010CB631100 and the Aeronautical Science Foundation of China under Projects 20167FV8010 are gratefully acknowledged.

Conflicts of Interest: The authors declare no conflicts of interest.

References

1. Lubineau, G.; Rahaman, A. A review of strategies for improving the degradation properties of laminated continuous-fiber/epoxy composites with carbon-based nanoreinforcements. *Carbon* **2012**, *50*, 2377–2395. [CrossRef]

2. Cicala, G.; Pergolizzi, E.; Piscopo, F.; Carbone, D.; Recca, G. Hybrid composites manufactured by resin infusion with a fully recyclable bioepoxy resin. *Composites Part B* **2018**, *132*, 69–76. [CrossRef]

3. Ma, S.; Li, T.; Liu, X.; Zhu, J. Research progress on bio-based thermosetting resins. *Polym. Int.* **2016**, *65*, 164–173. [CrossRef]

4. Chegdani, F.; Mansori, M.E.; Mezghani, S.; Montagne, A. Scale effect on tribo-mechanical behavior of vegetal fibers in reinforced bio-composite materials. Composites Science and Technology. *Compos. Sci. Technol.* **2017**, *150*, 87–94. [CrossRef]

5. Zhang, D.; Ye, L.; Deng, S.; Zhang, J.; Tang, Y.; Chen, Y. CF/EP composite laminates with carbon black and copper chloride for improved electrical conductivity and interlaminar fracture toughness. *Compos. Sci. Technol.* **2012**, *72*, 412–420. [CrossRef]

6. Kim, H.S.; Hahn, H.T. Graphite fiber composites interlayered with single-walled carbon nanotubes. *J. Compos. Mater.* **2011**, *45*, 1109–1120. [CrossRef]

7. Garcia, E.J.; Wardle, B.L.; Hart, A.J.; Yamamoto, N. Fabrication and multifunctional properties of a hybrid laminate with aligned carbon nanotubes grown In Situ. *Compos. Sci. Technol.* **2008**, *68*, 2034–2041. [CrossRef]

8. Sun, S.; Guo, M.; Yi, X. Phase separation morphology and mode II interlaminar fracture toughness of bismaleimide laminates toughened by thermoplastics with triphenylphosphine oxide group. *Sci. China Technol. Sci.* **2017**, *60*, 444–451. [CrossRef]

9. Sun, L.; Warren, G.L.; Davis, D.; Sue, H.-J. Nylon toughened epoxy/SWCNT composites. *J. Mater. Sci.* **2011**, *46*, 207–214. [CrossRef]

10. Groleau, M.R.; Shi, Y.-B.; Yee, A.F.; Bertram, J.L.; Sue, H.J.; Yang, P.C. Mode II fracture of composites interlayered with nylon particles. *Compos. Sci. Technol.* **1996**, *56*, 1223–1240. [CrossRef]

11. Yi, X.; Cheng, Q.; Liu, Z. Preform-based toughening technology for RTMable high-temperature aerospace composites. *Sci. China Technol. Sci.* **2012**, *55*, 2255–2263. [CrossRef]

12. Anand, A.; Kumar, N.; Harshe, R.; Joshi, M. Glass/epoxy structural composites with interleaved nylon 6/6 nanofibers. *J. Compos. Mater.* **2017**, *51*, 3291–3298. [CrossRef]

13. Yadav, S.N.; Kumar, V.; Verma, S.K. Fracture toughness behavior of carbon fibre epoxy composite with Kevlar reinforced interleave. *Mater. Sci. Eng. B* **2006**, *132*, 108–112. [CrossRef]

14. Khan, S.U.; Kim, J.-K. Improved interlaminar shear properties of multiscale carbon fiber composites with bucky paper interleaves made from carbon nanofibers. *Carbon* **2012**, *50*, 5265–5277. [CrossRef]

15. Palazzetti, R.; Yan, X.; Zucchelli, A. Influence of geometrical features of electrospun nylon6, 6 interleave on the CFRP laminates mechanical properties. *Polym. Compos.* **2014**, *35*, 137–150. [CrossRef]

16. Ramirez, V.A.; Hogg, P.J.; Sampson, W.W. The influence of the nonwoven veil architectures on interlaminar fracture toughness of the interleaved composites. *Compos. Sci. Technol.* **2015**, *110*, 103–110. [CrossRef]

17. Heijden, S.; Daelemans, L.; Meireman, T.; Baere, I.D.; Rahier, H.; Paepegem, W.V.; Clerck, K.D. Interlaminar toughening of resin transfer molded laminates by electrospun polycaprolactone structures: Effect of the interleave morphology. *Compos. Sci. Technol.* **2016**, *136*, 10–17. [CrossRef]

18. KuWata, M.; Hogg, P.J. Interlaminar toughness of interleaved CFRP using non-woven veils: Part 1. Mode-I testing. *Composites Part A* **2011**, *42*, 1551–1559. [CrossRef]

19. KuWata, M.; Hogg, P.J. Interlaminar toughness of interleaved CFRP using non-woven veils: Part 2. Mode-II testing. *Composites Part A* **2011**, *42*, 1560–1570. [CrossRef]

20. Beckermann, G.W.; Pickering, K.L. Mode I and Mode II interlaminar fracture toughness of composite laminates interleaved with electrospun nanofibre veils. *Composites Part A* **2015**, *72*, 11–21. [CrossRef]

21. Daelemans, L.; Heijden, S.; Baere, I.D.; Rahier, H.; Paepegem, W.V.; Clerck, K.D. Using aligned nanofibres for identifying the toughening micromechanisms in nanofibre interleaved laminates. *Compos. Sci. Technol.* **2016**, *124*, 17–26. [CrossRef]

22. Daelemans, L.; Heijden, S.; Baere, I.D.; Rahier, H.; Paepegem, W.V.; Clerck, K.D. Nanofibre bridging as a toughening mechanism in carbon/epoxy composite laminates interleaved with electrospun polyamide nanofibrous veils. *Compos. Sci. Technol.* **2015**, *117*, 244–256. [CrossRef]

23. Guo, M.; Yi, X.; Liu, G.; Liu, L. Simultaneously increasing the electrical conductivity and fracture toughness of carbon-fiber composites by using silver nanowires-loaded interleaves. *Compos. Sci. Technol.* **2014**, *97*, 27–33. [CrossRef]

24. Guo, M.; Yi, X. The production of tough, electrically carbon fiber composite laminates for use in airframe. *Carbon* **2013**, *58*, 241–244. [CrossRef]

25. Eskizeybek, V.; Yar, A.; Avcı, A. CNT-PAN hybrid nanofibrous mat interleaved carbon/epoxy laminates with improved Mode I interlaminar fracture toughness. *Compos. Sci. Technol.* **2018**, *157*, 30–39. [CrossRef]

26. Zhou, H.; Du, X.; Liu, H.-Y.; Zhou, H.; Zhang, Y.; Mai, Y.-W. Delamination toughening of carbon fiber/epoxy laminates by hierarchical carbon nanotube-short carbon fiber interleaves. *Compos. Sci. Technol.* **2017**, *140*, 46–53. [CrossRef]

27. Zheng, N.; Huang, Y.; Liu, H.-Y.; Gao, J.; Mai, Y.-W. Improvement of interlaminar fracture toughness in carbon fiber/epoxy composites with carbon nanotubes/polysulfone interleaves. *Compos. Sci. Technol.* **2017**, *140*, 8–15. [CrossRef]

28. Lee, S.-H.; Kim, H.; Hang, S.; Cheong, S.-K. Interlaminar fracture toughness of composite laminates with CNT-enhanced nonwoven carbon tissue interleave. *Compos. Sci. Technol.* **2012**, *73*, 1–8. [CrossRef]

29. Chen, C.Z.; Li, Y.; Yu, T. Interlaminar toughening in flax fiber-reinforced composites interleaved with carbon nanotube buckypaper. *J. Reinf. Plast. Compos.* **2014**, *33*, 1859–1868. [CrossRef]

Aerospace **2018**, *5*, 77

30. Ramamoorthy, S.K.; Skrifvars, M.; Persson, A. A review of natural fibers used in biocomposites: Plant, animal and regenerated cellulose fibers. *Polym. Rev.* **2015**, *55*, 107–162. [CrossRef]

31. Cesano, F.; Zaccone, M.; Armentano, I.; Cravanzola, S.; Muscuso, L.; Torre, L.; Kenny, J.M.; Monti, M.; Scarano, D. Relationship between morphology and electrical properties in PP/MWCNT composites: Processing-induced anisotropic percolation threshold. *Mater. Chem. Phys.* **2016**, *180*, 284–290. [CrossRef]

32. Ge, J.; Yao, H.-B.; Wang, X.; Ye, Y.-D.; Wang, J.-L.; Wu, Z.-Y.; Liu, J.-W.; Fan, F.-J.; Gao, H.-L.; Zhang, C.-L.; et al. Stretchable conductors based on silver nanowires: Improved performance through a binary network design. *Angew. Chem. Int. Ed.* **2013**, *52*, 1654–1659. [CrossRef] [PubMed]

33. Wu, C.; Fang, L.; Huang, X.; Jiang, P. Three-dimensional highly conductive graphene-silver nanowire hybrid foams for flexible and stretchable conductors. *ACS Appl. Mater. Interfaces* **2014**, *6*, 21026–21034. [CrossRef] [PubMed]

34. Haznedar, G.; Cravanzola, S.; Zanetti, M.; Scarano, D.; Zecchina, A.; Cesano, F. Graphite nanoplatelets and carbon nanotubes based polyethylene composites: Electrical conductivity and morphology. *Mater. Chem. Phys.* **2013**, *143*, 47–52. [CrossRef]

35. Cesano, F.; Rattalino, I.; Demarchi, D.; Bardelli, F.; Sanginario, A.; Gianturco, A.; Veca, A.; Viazzi, C.; Castelli, P.; Scarano, D.; et al. Structure and properties of metal-free conductive tracks on polyethylene/multiwalled carbon nanotube composites as obtained by laser stimulated percolation. *Carbon* **2013**, *61*, 63–71. [CrossRef]

Article

Sound Absorption Characterization of Natural Materials and Sandwich Structure Composites

Jichun Zhang [1], Yiou Shen [1,*], Bing Jiang [1] and Yan Li [1,2]

[1] School of Aerospace Engineering and Applied Mechanics, Tongji University, Shanghai 200092, China; zhangjichun@tongji.edu.cn (J.Z.); 18686380740@163.com (B.J.); liyan@tongji.edu.cn (Y.L.)
[2] Key Laboratory of Advanced Civil Engineering Materials, Ministry of Education, Tongji University, Shanghai 200092, China
* Correspondence: shenyiou@tongji.edu.cn; Tel.: +86-21-6598-5919

Received: 19 June 2018; Accepted: 5 July 2018; Published: 11 July 2018

Abstract: Natural fiber and wood are environmentally friendly materials with multiscale microstructures. The sound absorption performance of flax fiber and its reinforced composite, as well as balsa wood, were evaluated using the two-microphone transfer function technique with an impedance tube system. The microstructures of natural materials were studied through scanning electrical microscope in order to reveal their complex acoustical dissipation mechanisms. The sound absorption coefficients of flax fiber fabric were predicted using a double-porosity model, which showed relatively accurate results. The integrated natural materials sandwich structure was found to provide a superior sound absorption performance compared to the synthetic-materials-based sandwich structure composite due to the contribution of their multiscale structures to sound wave attenuation and energy dissipation. It was concluded that the natural-materials-based sandwich structure has the potential of being used as a sound absorption structure, especially at high frequency.

Keywords: plant fiber; balsa; sound absorption; microstructures; sandwich structures

1. Introduction

Over the last couple of decades, there has been an increasing demand for improving the interior noise and indoor noise of aircraft, railway, automobile, and building compartments. Interior noise impairs people's health and causes a decrease in passenger comfort and active safety performance [1]. The use of sound-absorbing materials is one of the present effective noise control technologies. The conventional industrial sound absorption and insulation materials are mineral fibers, foam, and their composites. However, the usage of such materials is expensive, energy-consumptive, and adds more weight to the structures, which impacts their structural integrity.

Sandwich structure composites have been extensively employed in aerospace, transportation, construction, and new energy fields as load-bearing components due to their high strength-to-weight ratio. Sandwich structure composites can be designable and multifunctional, and more functions can be provided besides bearing with proper design, such as impact tolerance, thermal isolation, radiation resistance, acoustic absorption, etc. Nowadays, growing attention has focused on environmentally friendly materials that can be recycled or need less energy for production and contribute minimum greenhouse gas emissions. Utilizing natural materials as face sheets or core materials for sandwich structures would provide superior load-bearing and fatigue properties. Moreover, this kind of structure provides good sound absorption properties compared to the conventional metallic and laminated composite structural members [2]. These natural materials and sandwich structures have the potential of replacing synthetic materials in applications with demands of weight constraints or environmental friendliness, such as interior panels or floors in the aerospace industry, high-speed trains, and automobiles.

Zheng and Li [3] found using plant-fiber-reinforced composite (PFRC) as skins of sandwich structures processes better sound absorption properties than using glass-fiber-reinforced composite (GFRC) as skins. The damping properties of flax-fiber-reinforced composite (FFRC) face sheets and GFRC face-sheet-based sandwich structures were compared by Petrone et al. [4,5]. Their research results indicated that FFRC face sheet sandwich panels have outperformed energy dissipation ability compared to the synthetic fiber face sheet ones. Sarginis et al. [6] investigated the sound and vibration damping characteristics of balsa-core-based sandwich beams. They found that the coincidence frequency of sandwich beams with natural fiber facing sheet and balsa core was tripled compared to a synthetic-core-based sandwich beam. Plant fiber is a kind of natural material, and it has been widely used as reinforcement in composite structures for the last decades due to their rich source, low price, specific strength, high specific modulus, and so on. Plant fiber has a complex, multiscale, hollow structure compared with synthetic fiber [7,8]. For example, flax fiber is extracted from the stalk of the plant, and the characteristics of its natural growth result in its multiscale structure, which can be simplified at two scales. At the mesoscopic scale, a bundle contains 10 to 40 elementary fibers (12–16 μm in diameter, 2–5 cm in length) which are linked together mainly by pectin. At the microscopic scale, different from the solid structure of synthetic fiber, each elementary fiber is made up of concentric cell walls called the primary cell wall (0.2 μm) and the secondary cell wall. The secondary cell wall is composed of three layers, namely the S1 layer (0.08–0.2 μm), the S2 layer (1–10 μm), and the S3 layer (0.1 μm). Most of the cellulose microfibrils are located in the S2 layer, which is a type of polymer and possesses viscoelastic properties. The S2 layer of flax fiber consists of numerous cellulose microfibril sublayers ($L_1, L_2 \ldots L_n$) which run parallel to each other and form a microfibril angle with the fiber direction. At the center of the elementary fiber, there is a 2–5-μm diameter small open channel called lumen, as revealed in Figure 1a. It can be seen from Figure 1b that hundreds of elementary fibers twist together to form a continuous yarn, and a gap of dozens of microns is between each elementary fiber. The unique microstructure brings many distinctive advantages for plant fiber, such as superior energy absorption [9], acoustic [10] and damping [11] performances, etc. Natural fiber was found to be a good acoustically absorbent material due to the hollow structures which efficiently convert acoustic energy into mechanical and heat energy [12]. The investigations on the acoustic absorption properties of natural-fibrous-material-reinforced composites are limited; however, several scholars have experimentally studied the sound absorption coefficients of plant-fiber-reinforced composites (PFRC) and found their sound absorption coefficients (SAC) were much higher than those of the synthetic-fiber-reinforced composites at 100–2000 Hz frequency level [12,13]. Yang and Li [14] also found that the Garai–Pompoli model and Delany–Bazley model have good agreement with the experimental results. However, the acoustical dissipation mechanisms of plant fiber materials were not plenarily analyzed in these studies and the multiscale hollow structure of plant fibers was not reflected by these two models. Therefore, the multiscale structure of plant fiber should be considered during sound absorption mechanisms analysis as well as the acoustic properties prediction. Olny and Boutin [15] theoretically studied the acoustic wave propagation in media with two interconnected networks of pores of very different characteristic sizes. In this study, flax fiber yarn was considered as a double-porosity media, and this double-porosity model was employed to predict the acoustic properties of flax fiber.

The relatively high mechanical properties and low density of balsa make it attractive for cores in sandwich panels, and more importantly, sustainable. To date, there are no engineered materials suitable for sandwich panel cores with a similar combination of mechanical properties and low density [16]. Balsa wood can be considered as a kind of cellular material; it is composed of different size cell fibers, rays, and vessels, and there are also many micropores connecting each grain. It can be seen from Figure 2a,b that the fibers are long prismatic cells, resembling a hexagon in cross-section, ranging from 20 to 40 μm in diameter. The rays are brick-like parenchyma cells ranging from 20 to 40 μm in cross-section. The vessels are long, tubular structures that run axially along the trunk of the tree, and the diameter of the vessels is 200–350 μm. Tiny micropores with a diameter of a few microns generally exist on the cell wall as seen in Figure 2c, which make it a kind of semi-open cellular structure. The thickness of the double cell wall is about 0.8 to 3 μm in fibers as shown in Figure 2d. Balsa wood also possesses a multiscale structure, and the cell wall is similar to that of natural fiber, which consists of a primary layer and three secondary layers, the S1, S2, and S3, and the S2 layer is generally the thickest layer [17]. The cell wall of wood is a composite structure in which cellulose microfibrils are reinforced with hemicelluloses and lignin. Therefore, this structure lowers the density and has a competitive mechanical performance with commercial close-cell foam. Many researchers have investigated the mechanical performance of balsa [18,19] and balsa-based sandwich structures [20,21], and they found the balsa core sandwiches were stiffer than traditional polymer-based sandwiches and have better energy absorption capacity. Only a few researchers have focused on the acoustic absorption properties of this natural cellular-material-based sandwich structure, and they believed that balsa, with a high volume of pores, also has great potential as noise absorption material [6].

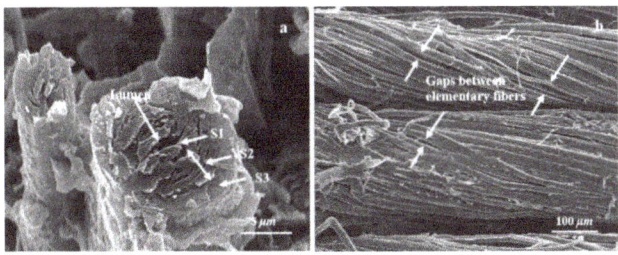

Figure 1. (**a**) Cross-section of an elementary flax fiber and (**b**) flax fiber yarns.

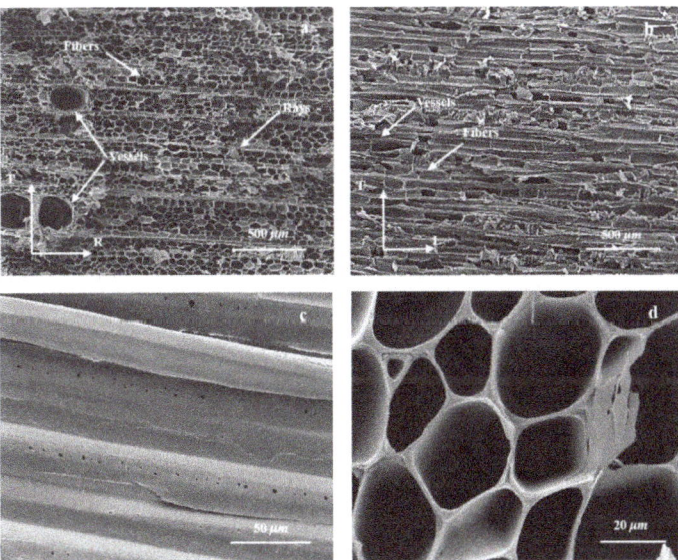

Figure 2. SEM micrographs of balsa wood (**a**) cross-section; (**b**) longitudinal view; (**c**) micropores on fibers; and (**d**) multilayer cell wall. L, R, and T refer to the longitudinal, radial, and tangential axes of symmetry.

Considering the complicated multiscale structure of natural materials, different acoustic absorption mechanisms may be found for the natural materials and their composite structures compared to the synthetic fiber and reinforced composites. However, the above studies have not focused on this point, thus the sound absorption mechanism of natural materials and its composite structures is still unclear. In this paper, the sound absorption performance of flax fiber, reinforced composite, balsa wood, and the integrated sandwich structure composites were evaluated using the impedance tube method. The acoustical dissipation mechanisms were analyzed on the point of view of their unique multiscale microstructures. With a purpose of comparison, commonly used sandwich composites with glass fiber face sheets and polyethylene terephthalate (PET) foam core were also investigated.

2. Materials and Experiments

2.1. Materials

Flax-fiber-fabric-reinforced epoxy composite was used as skins and balsa wood was used as the core material for Flax-Balsa sandwich structures. Likewise, Flax-PET and Glass-PET sandwich structures were prepared for comparison. The unidirectional flax fabric was provided by Lineo Co. Ltd., Saint Martin Du Tilleul, France, and the unidirectional glass fabric was provided by Zhejiang Mengtai Reinforced Composite Material Co. Ltd., Jiaxing, China. These two fabrics had similar areal density, however, the thickness of the glass fiber fabric was only half of the flax fiber fabric due to the density reason. The NPEL-128 epoxy was supplied by Shanghai Zhongsi Industry Co. Ltd., Shanghai, China. The balsa wood and the AIREX® T92.100 close-cell PET foam were supplied by Zhuhai Dechi Technology Co. Ltd., Zhuhai, China and 3A Composites (China) Ltd., Shanghai, China, respectively. The detail parameters of the flax and glass fiber fabric used for fabricating the sandwich structure skins are listed in Table 1.

Balsa wood and PET foam panel with a thickness of 10 mm were prepared as core materials to fabricate the sandwich structure. The basic parameters of these two types of core material are shown in Table 2. Balsa wood and PET foam with similar density and porosity were selected in this study for comparison.

Table 1. The details of fiber fabrics used in manufacturing sandwich structures skins.

Materials	Fiber Density (g/cm^3)	Mean Fiber Diameter (μm)	Mean Yarn Diameter (μm)	Areal Density (g/m^2)	Mean Thickness (mm)	Specific Strength (MPa/g/cm^3)	Specific Modulus (GPa/g/cm^3)
Flax fabric	1.5	15	200	213.3	0.2	1034	55
Glass fabric	2.55	11	100	207.5	0.1	980	30

Table 2. The parameters of core materials used in manufacturing sandwich structures.

Materials	Density (kg/m^3)	Porosity	Cellular Diameter (μm)	Axial Compression Modulus (MPa)	Transverse Compression Modulus (MPa)
Balsa wood	125	0.920	35–200	280–320	30–80
PET foam	112	0.933	500–700	90	90

2.2. Fabrication

The epoxy resin, curing agent, and accelerating agent were mixed at the weight ratio of 100:80:1. Initially, 12 layers of flax fiber fabrics and 24 layers of glass fiber fabrics were impregnated with the resin system and laid inside of a steel mold in stacking sequence of $[0/90/0]_{2s}$ and $[0_2/90_2/0_2]_{2s}$, respectively. In addition, 6 layers of impregnated flax fiber fabrics were laid on the top and bottom of the core material with stacking sequence of $[0/90/0]_s$ in order to fabricate flax-fiber-reinforced composite (FFRC) skin-based sandwich panels. A similar process was used for GFRP skin-based sandwich panels (12 layers on each side, $[0_2/90_2/0_2]_s$). The resulting laminates and sandwich prepreg systems were then respectively placed in a hot press machine under pressure of 1 MPa and temperature of 120 °C for 2 h for complete curing. Two types of composite laminate, being flax-fiber-reinforced composite (FFRC) and glass-fiber-reinforced composite (GFRC), with mean thickness of 4 mm and fiber volume fraction of 50% were manufactured. Three types of sandwich structures, Flax-Balsa, Flax-PET, and Glass-PET, with 2 mm thickness and 50% fiber volume fraction of composite skin on each side were manufactured, in which Flax-Balsa and Flax-PET denote sandwich structures based on FFRC skins and balsa or PET foam cores, respectively. The thickness of the balsa panel was on the tangential direction of the balsa wood. The Glass-PET denotes a sandwich structure with GFRC skins and PET foam core, as illustrated in Figure 3.

Figure 3. Schematic diagram of sandwich structures.

2.3. Acoustic Property Measurement

Sound absorption measurement was performed using the impedance tube method. The sound absorption coefficient (SAC) of sandwich composites was measured through a transfer function technique with the impedance tube facilities manufactured by BSWA Technology Co. Ltd., Beijing, China. The SAC is defined as the ratio of the absorbed sound energy and the incident sound energy. The measurements were conducted according to ISO10534-2 standard at medium to high frequencies from 250 to 10,000 Hz at 25 °C and 60% relative humidity, and at least three specimens were tested for each group. The measurements were composed of SW422, SW477, and SW499 impedance tubes with diameters of 100, 30, and 16 mm for measuring frequencies from 250–2000, 800–6300, and 2500–10,000 Hz, respectively. Samples were backed up by a rigid wall which reflected all the incoming sound energy. Two microphones were mounted at the wall of the tube to measure the sound pressure. Therefore, if the incident sound energy was known and the transmission sound energy could be measured with the aid of the transfer function, the sound energy absorbed by the materials could be obtained, as illustrated in Figure 4.

The SACs of 12 layers of flax fiber fabric and 24 layers of glass fiber fabric with same thickness were initially measured at frequency ranging from 250 to 4000 Hz. The composite laminates and sandwich structures were carefully cut to the required size and wrapped with Teflon tape provided by 3M Company, Maplewood, MN, USA to prevent air leaks between the sample and tube. All the samples were put in the oven at 120 °C for 4 h to remove moisture.

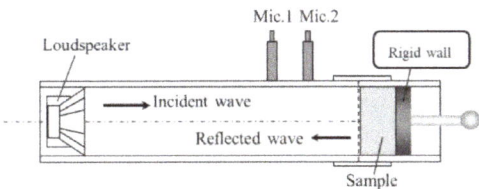

Figure 4. Instrumentation for the transfer function method of measuring sound absorption coefficients.

3. Results and Discussions

3.1. Sound Absorption Performance of Flax Fiber and Its Reinforced Composite

The tested sound absorption coefficients of unidirectional flax fiber fabric and glass fiber fabric are presented in Figure 5 as a function of frequency. The results show that in the test frequency range of 250–4000 Hz, flax fiber fabric has superior acoustic absorption ability compared to glass fiber fabric. It can be seen from Figure 5 that the SACs of flax fiber fabric were higher than 0.5 when the sound wave frequency was over 1000 Hz, which means over half of the incident sound energy was absorbed by the fabric. In this study, the critical frequency of flax fiber fabric was 3200 Hz, which means the SAC reached a peak value of 0.96 at 3200 Hz and showed a relatively stable sound absorption ability afterwards. Clearly, the maximum SAC of glass is much lower than that of the flax fiber, only 0.58 in the test frequency range, and the critical frequency is higher than that of the flax fiber (>4000 Hz). This means flax fibers process not only outstanding sound absorption capability but also wider sound absorbing frequency range.

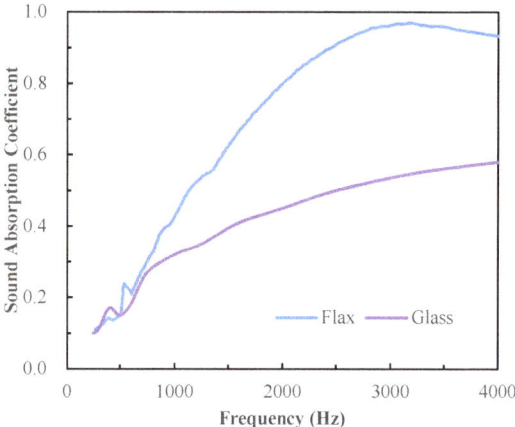

Figure 5. The experimental sound absorption coefficients of unidirectional flax and glass fiber fabrics with frequency.

It is well known that the mechanism of sound attenuation for fibrous materials is due to the interaction between acoustic waves and fiber assemblies, and the resonant frequency of sound wave absorption could be mainly attributed to the vibration of fibrous materials [12]. For natural fibrous material, the acoustic absorption mechanism is more complicated compared to the synthetic fibrous material due to its complex microstructure. As the mechanism illustrated in Figure 6 shows, the hollow and multiscale structure of flax fiber acts on the reduction and dissipation of sound energy in several manners. Firstly, the incident sound wave arouses the vibration of air molecules between elementary fibers and within the lumens, which causes interaction between air and the fiber cell wall. The resulting viscous resistance transfers this part of acoustic energy to thermal energy. Meanwhile, thermal exchange between neighboring particles of fiber also results in acoustic attenuation. Most importantly, differing from the solid structure of glass fiber, vibration and friction of microscale or even nanoscale microfibrils in elementary flax fiber also have an effect on the extra dissipation of sound energy besides vibration and friction between elementary fibers, leading to a more efficient conversion from sound energy to thermal energy. It can be seen from Figure 1a that there existed a large number of gaps between adjacent S2 sublayers in some regions. In this case, interfacial friction between adjacent sublayers may also dissipate energy. Moreover, the large fiber yarn diameter of flax fiber fabric causes bigger interspace between adjacent fibers, which leads to the better mobility of air in fabric and smaller surface impedance of the composite. Therefore, the sound wave can easily transmit into the flax fiber fabric and be absorbed or dissipated, consequently enhancing the sound absorption of the materials. This explains the high sound absorption coefficients of flax fiber in Figure 5.

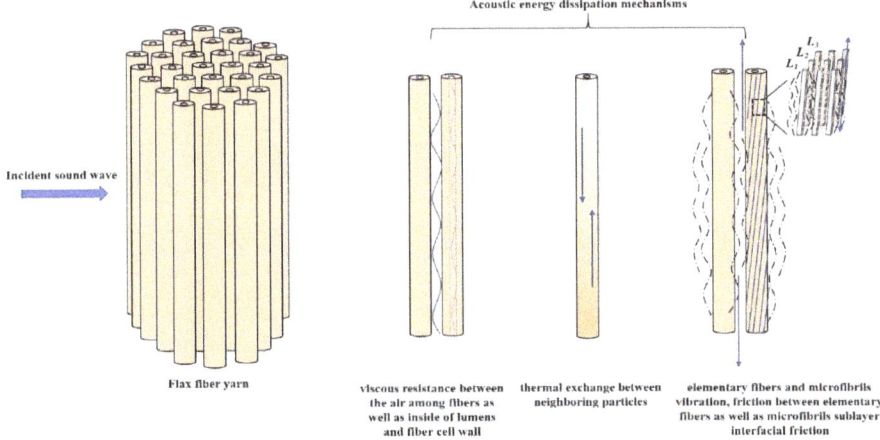

Figure 6. Acoustic energy dissipation mechanisms of flax fiber yarns.

The superior sound absorbing ability of plant fiber is essentially attributed to its multiscale micromorphology. However, the multiscale hollow structure of plant fibers was not reflected by the Garai–Pompoli model or Champoux–Allard model in predicting the SACs of plant fibers. In this study, the lumen of flax fiber was considered in predicting its acoustic properties. Firstly, an idealized flax elementary fiber was assumed as a 15-μm diameter tubular cell with a 5-μm diameter lumen inside. The effective density ρ_{eq} and the effective modulus K_{eq} of unidirectional flax fiber fabric were predicted according to the double-porosity model developed by Olny and Boutin [15] as shown in Equations (1) and (2):

$$\rho_{eq}(\omega) = \left(\frac{1}{\rho_p(\omega)} + \frac{1-\varnothing_p}{\rho_m(\omega)} \right)^{-1} \tag{1}$$

$$K_{eq}(\omega) = \left(\frac{1}{K_p(\omega)} + \frac{(1-\varnothing_p)F_d(\omega,\omega_d)}{K_m(\omega)} \right)^{-1}, \quad \omega_d \approx \frac{P_0}{\varnothing_m \sigma_m l_p^2} \tag{2}$$

where subscripts 'p' and 'm' present the mesoscopic scale (between fibers) and microscopic scale (within the fiber). l_p and l_m present the characteristic length in these two scales. $\rho_p(\omega)$, $\rho_m(\omega)$, and $K_p(\omega)$ were obtained by plugging the basic mesoscopic and microscopic scales acoustic parameters into the Champoux–Allard model, Equations (3) and (4). F_d is a pressure diffusion function. In this study, the test frequency $\omega < \omega_d$, $F_d \approx 1$ and voids at both scales participated in the attenuation of acoustic energy; the sound pressure between fibers and lumens are the same.

$$\rho_{eq}(\omega) = \frac{\rho_0 \alpha_\infty}{\varnothing} \left[1 - i\frac{\sigma\varnothing}{\omega\rho_0\alpha_\infty} \sqrt{1 + i\omega\rho_0\eta \left(\frac{2\alpha_\infty}{\sigma\varnothing\Lambda} \right)^2} \right] \tag{3}$$

$$\frac{1}{K_{eq}}(\omega) = \frac{\varnothing}{\gamma P_0} \left\{ \gamma - (\gamma-1) \left[1 - i\frac{8\eta}{\omega\rho_0 P_r \Lambda'^2} \sqrt{1 + \frac{i\omega\rho_0 P_r}{\eta} \left(\frac{\Lambda'}{4} \right)^2} \right]^{-1} \right\} \tag{4}$$

where ω is the angular frequency, $\omega = 2\pi f$, f is the incident sound wave frequency, $\varnothing$ is the porosity, and α_∞ is the tortuosity, which characterizes the buckling of channels in the material. For ideal fiber materials, $\alpha_\infty = 1.34$. Λ and Λ' are viscous characteristic length and thermal characteristic dimension, which characterize the average diameter of unit length for viscous loss and thermal loss, respectively. Λ and Λ' are defined as $\Lambda = 1/2\pi rL$, $\Lambda' = 2\Lambda$, in which $L = 4\rho f/\pi a^2 \rho_b$, $r = a/2$, a is fiber diameter and

ρ_b is the fabric density. ρ_0 and γ are air density and specific heat ratio, respectively, and P_0 is the atmospheric pressure. P_r is the Prandtl number, defined as $P_r = C_p/k$, where C_p is isobaric heat capacity, μ is viscosity, and k is thermal conductivity. The Prandtl number characterizes the effect of the physical characteristic of air on the heat transfer process. The porosity can be obtained by Equation (5)

$$\varnothing = 1 - \frac{\rho_b}{\rho_f} \tag{5}$$

where ρ_f is the fiber density.

The resulting ρ_{eq} and K_{eq} were then substituted into Equation (6) to get the acoustic characteristic impedance, Z_{eq}, and wave number, k_{eq}, of flax fiber fabric. The sound absorption coefficients can then be obtained using Equation (9)

$$Z_{eq} = \sqrt{\rho_{eq}K_{eq}}, \quad k_{eq} = \omega\sqrt{\frac{\rho_{eq}}{K_{eq}}} \tag{6}$$

$$\alpha = 1 - \left|\frac{Z_n - 1}{Z_n + 1}\right|^2, \quad Z_n = -i\frac{Z_{eq}}{\rho_0 c_0}\coth(k_{eq}t) \tag{7}$$

where c_0 is the sound velocity in air, t is the thickness of material. It can be observed from Equations (3)–(7) that the acoustic parameters of material are related to its physical parameters, including fiber diameter, fiber density, fabric density, and fabric thickness. It indicates that these physical parameters of material play a critical role on the sound absorption properties.

It can be found from Figure 7 that the double-porosity model showed more accuracy predicting results compared to the Champoux–Allard model [22], and the critical frequency of flax fiber fabric was accurately estimated. This indicates that the main factors affecting the sound absorption properties of plant fiber materials are not only fiber diameter, density, and fabric density, but the lumen diameter is also an important parameter. The double-porosity model can be used to predict the sound absorption coefficients of plant fibers effectively.

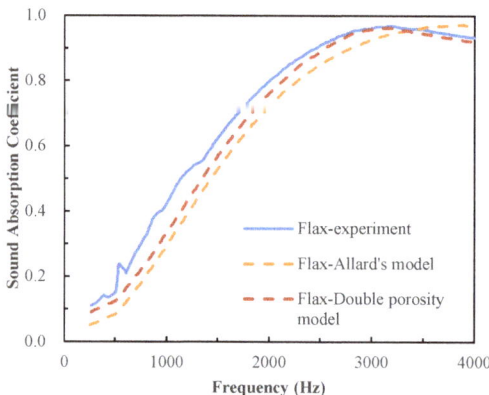

Figure 7. Comparison of experimental and theoretical sound absorption coefficients of unidirectional flax fiber fabrics with frequency.

Figure 8 compares the experimental results on SACs of flax fiber and glass-fiber-reinforced composites. FFRC laminate provides superior sound absorption capacity to GFRC especially at higher frequency, as expected. However, the SACs of the composites are obviously much lower than the fiber fabric and the effective sound absorption frequency range (SACs value over 0.2) was reduced. This is due to the reason that when an incident sound wave reaches the surface of material, partial sound energy is reflected. The reflected energy depends on the surface acoustic impedance, Z, which can be described using Equation (8).

$$Z = \sqrt{E \cdot \rho} \tag{8}$$

where E is the modulus of the materials (sound wave incident direction) and ρ is the density of the materials. The sound wave can be easily reflected when Z is high and more sound energy is reflected. In contrast, a substantial part of sound energy enters into the materials when Z is low. Apparently, more sound energy was reflected by the composite and this limited its sound absorption ability. Furthermore, the microstructure of the material structure had a major effect on energy dissipation at high frequency due to the very short wavelength of sound wave, which made it easily eliminated when transmitted through the media. The SACs of pure epoxy are very low, generally below 0.2. Thus, the sound absorption performance of composite laminate is mainly dependent on the sound absorption characteristics of fiber. Previous work [23] has found that the multilayer structure of flax fiber provided more internal interfaces (primary cell wall, S1, S2, and S3 layers of the secondary cell wall, and S2 layer consist of numerous cellulose microfibril sublayers) and paths inside of flax fiber to dissipate energy compare to the solid synthetic fiber. Therefore, the damping property of flax fibers was much better and resulted in a higher damping ratio for its reinforced composite. This indicates that the FFPC possesses higher energy dissipation and acoustic–thermal conversion capacity, and it also evidences its better acoustic absorption performance than that of GFRP. Prabhakaran et al. [13] also found FFRC process better SACs and damping factors compared to GFRC.

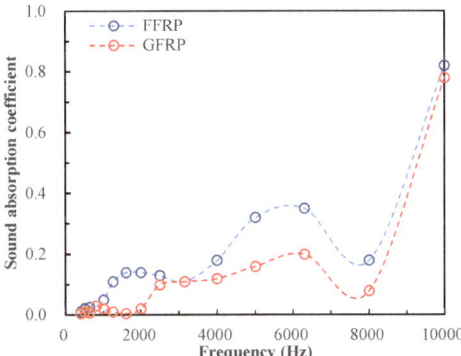

Figure 8. The experimental sound absorption coefficients of FFRC and GFRC with frequency.

3.2. Sound Absorption Performance of Balsa Wood

Figure 9 illustrates the SACs of balsa and PET foam, which indicate that these two types of core materials possess similar sound absorption ability along the test frequencies. The peak value of SACs below 10,000 Hz for these two materials are 0.4 and 0.42, respectively. The maximum SAC of these two porous materials at 10,000 Hz are also very close, being 0.85 and 0.87. The similar density and porosity of these two materials lead to their essentially identical trend on the sound absorption behavior. The cell wall of balsa is composed of polymer materials, and the visco-inertial and thermal damping dominates the sound absorption behavior for polymeric foams. In low frequency stage, thermal exchange is dominated in energy dissipation, whereas in high frequency, viscous resistance is the main energy consuming method. The loss of sound energy is mainly due to the contribution of the porous sound absorption mechanism and resonance absorption mechanism. The air molecules periodically vibrate under the influence of sound waves; meanwhile, the friction between the air molecules and cell wall cause frictional heat. On the other hand, the compression and expanding deformation of air inside of the cell occurs; this transfers partial sound energy into thermal energy. In terms of balsa, vibration and friction of cellulose microfibrils in the cell wall and interfacial friction between adjacent sublayers caused sound energy dissipation, which is similar to the flax fiber. In addition, the micropores in the cell wall of balsa consumed sound energy due to this complex inner geometry providing a tortuous sound path through the foam. It can be found that natural material has a complicated acoustical energy dissipation mechanism due to its natural growth multiscale structure which reveals a satisfactory sound absorption performance.

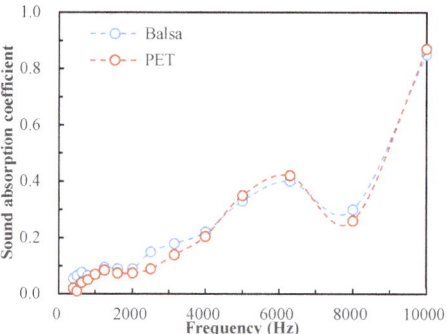

Figure 9. The experimental sound absorption coefficients of balsa and PET foam with frequency.

3.3. Sound Absorption Performance of Sandwich Structures

An effective way to improve the acoustic absorption performance of composite laminate is inducing a porous material as the core material to cooperate a sandwich structure. Meanwhile, the sound wave can be reflected repeatedly by the stiff composite skins back to the interior of the cellular core material for energy exchange and dissipation. Figure 10 shows the variation of SACs with frequency of Flax-Balsa, Flax-PET, and Glass-PET sandwich structures and their corresponding core materials. It can be seen from Figure 10 that the SACs increase with frequency for all the samples, reaching at least 0.8 at 10,000 Hz. For flax-fiber-reinforced composite skin-based sandwich structures, the SACs of the sandwich panel are higher than those of the core materials in all frequency levels. The SACs are clearly improved for the balsa if cooperated with FFRC skins from 1600 Hz. The sound absorption peak of the Flax-Balsa sandwich panel is 0.65, which is approximately 63% and 86% enhancement compared to that of the balsa and FFRC, respectively. On the other hand, the SACs of PET foam have been distinctly enhanced over 40% at high frequency from 4000 Hz when cooperated with FFRC skins. It can be concluded that flax fiber skin-based sandwich structures provide better sound absorption performance than their corresponding core materials at a large range of frequency band, and the composite skins definitely play a significant role. The dashed line represents the effective sound absorption peak width (SACs value over 0.2) which has been enlarged from 4000–8000 Hz (balsa) to 3150–8000 Hz (Flax-Balsa). Nevertheless, the SACs of the sandwich structure with flax-fiber-reinforced skins are overall higher compared to the sandwich structures with glass fiber skins, especially at higher frequency due to the high surface acoustic impedance and low sound absorption ability of GFRC skins. Thus, choosing the appropriate fiber type for composite skins will improve the acoustic performance of sandwich structures effectively.

It is worth noting that when cooperated with flax fiber reinforced skins, the SACs of Flax-Balsa sandwiches are clearly higher than those of Flax-PET despite their core materials possessing similar acoustic behavior with the same skins. This is due to the fact that PET foam has larger cell size ranging from 500 to 700 μm, whereas the cell size of balsa ranges from 35 to 200 μm. The epoxy resin filled the layer of cells that contacted the skin during fabrication. After curing, the solid layer of epoxy resin in Flax-PET sandwich structures was much thicker than that of Flax-Balsa; this limited the sound wave entering into the core materials and resulted in its relatively lower SACs. It is clear from Figure 10 that the Flax-Balsa sandwich structure has the best sound absorption behavior over most of the frequency band compared to the other two sandwich structures and their core materials. The superior acoustic performance of this environmentally friendly sandwich structure plays a great role in their unique multiscale microstructure advantages.

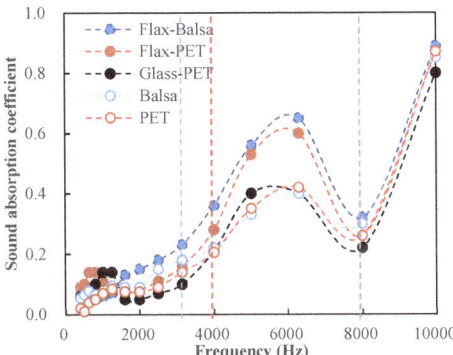

Figure 10. The experimental sound absorption coefficients of sandwich structures and their corresponding core materials.

Figure 11 illustrates the sound absorption mechanism of natural-material-based sandwich structure composites. When sound energy (E_i) vertically incidents from one side of the structure, partial sound energy is reflected (E_r) by the surface of the composite skin. The incident sound energy is initially absorbed (E_{a1}) by the FFRC skin through viscous resistance, thermal exchange, and damping of fibers or microfibrils and multilayer cell walls. Partial energy transmits through top skin (E_{t1}) and some energy reflects by the interface of skin and core (E_{r1}). Balsa core absorbs sound energy (E_{a2}) mainly through visco-inertial and thermal damping, multilayer cell wall damping, and air flow through micropores also causing sound energy dissipation. The residual sound energy was reflected (E_{r2}), absorbed (E_{a3}), and transmitted by the bottom skin (E_{t1}).

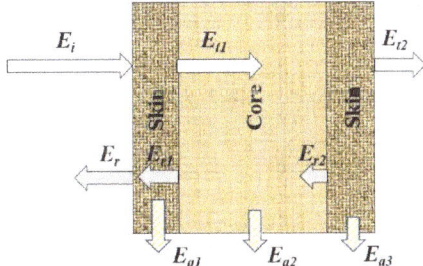

Figure 11. Schematic diagram of sound energy absorption mechanism for sandwich structure.

4. Conclusions

Sound absorption properties of natural materials and their integrated sandwich structure were characterized and compared with synthetic materials. It was found that the multiscale structure of natural materials was the reason for their outstanding sound energy absorption performance and complicated energy dissipation mechanism. The double-porosity model was applied for predicting the sound absorption coefficients of plant fiber yarns and it showed more accurate results compared to the

Champoux–Allard model. This suggested that the unique multiscaled hollow structure has an effect on the acoustic behavior of plant fiber. The multilayer sandwich structure is conducive to effective and gradual dissipation of sound energy compared to the individual core material or composite laminates, especially at high frequency. With proper design, the natural-materials-based sandwich structure has the potential for being utilized as a load-bearing and sound-absorbing multifunctional structure, especially at high frequency, which would be very beneficial for aeronautical applications due to weight restriction and the high sound frequency service environment in order to increase the comfort of the passengers.

Author Contributions: Y.S. and Y.L. conceived and guided the project and study; J.Z. was responsible for the manufacturing of the sandwich structures and preparing samples; J.Z. and B.J. processed the experimental data and performed the calculation; J.Z. prepared the writing-original draft and Y.S. made the writing-review & editing.

Funding: This paper was supported by National Natural Science Foundation (11302151) and the Fundamental Research Funds for the Central Universities.

Acknowledgments: The authors are thankful to Lineo Co. Ltd. (Saint Martin Du Tilleul, France) and BSWA Technology Co. Ltd. (Beijing, China) for supplying the natural flax fibers and the impedance tube facilities for measuring SACs respectively.

Conflicts of Interest: The authors declare no conflict of interest.

References

1. Wilby, J.F. Aircraft interior noise. *J. Sound Vib.* **1996**, *190*, 545–564. [CrossRef]
2. Liu, J.L.; Bao, W.; Shi, L.; Zuo, B.Q.; Gao, W.D. General regression neural network for prediction of sound absorption coefficients of sandwich structure nonwoven absorbers. *Appl. Acoust.* **2014**, *76*, 128–137. [CrossRef]
3. Zheng, Z.Y.; Li, Y.; Yang, W.D. Absorption properties of natural fiber-reinforced sandwich structures based on the fabric structures. *J. Reinf. Plast. Compos.* **2013**, *32*, 1561–1568. [CrossRef]
4. Petrone, G.; D'Alessandro, V.; Franco, F.; De Rosa, S. Damping evaluation on eco-friendly sandwich panels through reverberation time (RT 60) measurements. *J. Vib. Control* **2015**, *21*, 3328–3338. [CrossRef]
5. Petrone, G.; D'Alessandro, V.; Franco, F.; Mace, B.; De Rosa, S. Modal characterisation of recyclable foam sandwich panels. *Compos. Struct.* **2014**, *113*, 362–368. [CrossRef]
6. Sargianis, J.J.; Kim, H.I.; Andres, E.; Suhr, J. Sound and vibration damping characteristics in natural material based sandwich composites. *Compos. Struct.* **2013**, *96*, 538–544. [CrossRef]
7. Fratzl, P.; Weinkamer, R. Nature's hierarchical materials. *Prog. Mater. Sci.* **2007**, *52*, 1263–1334. [CrossRef]
8. Charlet, K.; Jernot, J.; Eve, S. Multi-scale morphological characterisation of flax: From the stem to the fibrils. *Carbohydr. Polym.* **2010**, *82*, 54–61. [CrossRef]
9. Meredith, J.; Ebsworth, R.; Coles, S.R.; Wood, B.M.; Kirwan, K. Natural fibre composite energy absorption structures. *Compos. Sci. Technol.* **2012**, *72*, 211–217. [CrossRef]
10. Berardi, U.; Iannace, G. Acoustic characterization of natural fibers for sound absorption application. *Build. Environ.* **2015**, *94*, 840–852. [CrossRef]
11. Duc, F.; Bourban, P.E.; Plummer, C.J.G.; Månson, J.A.E. Damping of thermoset and thermoplastic flax fibre composites. *Compos. Part A-Appl. Sci.* **2014**, *64*, 115–123. [CrossRef]
12. Tang, X.; Yan, X. Acoustic energy absorption properties of fibrous materials: A review. *Compos. Part A-Appl. Sci.* **2017**, *101*, 360–380. [CrossRef]
13. Prabhakaran, S.; Krishnaraj, V.; Senthil kumar, M.; Zitoune, R. Sound and vibration damping properties of flax fiber reinforced composites. *Procedia Eng.* **2014**, *97*, 573–581. [CrossRef]
14. Yang, W.; Li, Y. Sound absorption performance of natural fibers and their composites. *Sci. China Technol. Sci.* **2012**, *55*, 2278–2283. [CrossRef]
15. Olny, X.; Boutin, C. Acoustic wave propagation in double porosity media. *J. Acoust. Soc. Am.* **2003**, *114*, 73–89. [CrossRef] [PubMed]
16. Damian, B. An introduction to core materials. *Reinf. Plast.* **2014**, *58*, 32–37.
17. Borrega, M.; Ahvenainen, P.; Serimaa, R.; Gibson, L. Composition and structure of balsa (*Ochroma pyramidale*) wood. *Wood Sci. Technol.* **2015**, *49*, 403–420. [CrossRef]

18. Borrega, M.; Gibson, L.J. Mechanics of balsa (*Ochroma pyramidale*) wood. *Mech. Mater.* **2015**, *84*, 75–90. [CrossRef]

19. Silva, A.D.; Kyriakides, S. Compressive response and failure of balsa wood. *Int. J. Solids Struct.* **2007**, *44*, 8685–8717. [CrossRef]

20. Atas, C.; Sevim, C. On the impact response of sandwich composites with cores of balsa wood and PVC foam. *Compos. Struct.* **2010**, *93*, 40–48. [CrossRef]

21. Osei-Antwi, M.; Castro, M.; Vassilopoulos, A.P.; Keller, T. Fracture in complex balsa cores of fiber-reinforced polymer sandwich structures. *Constr. Build. Mater.* **2014**, *71*, 194–201. [CrossRef]

22. Champoux, Y.; Allard, J. Dynamic tortuosity and bulk modulus in air-saturated porous media. *J. Appl. Phys.* **1991**, *70*, 1975–1979. [CrossRef]

23. Li, Y.; Cai, S.; Huang, X. Multi-scaled enhancement of damping property for carbon fiber reinforced composites. *Compos. Sci. Technol.* **2017**, *143*, 89–97. [CrossRef]

Article

Development of Bio-Sourced Epoxies for Bio-Composites

Xiao-Su Yi [1,2,*], Xvfeng Zhang [2,†], Fangbo Ding [3,†] and Jianfeng Tong [2,†]

1   Advanced Materials and Composites Department, Faculty of Science & Engineering,
    University of Nottingham (UNNC), Ningbo 315100, China
2   AVIC Composite Corporation Ltd. (ACC), Beijing 101300, China; 010xufeng@sina.com (X.Z.);
    broadtjf@sina.com (J.T.)
3   AVIC XAC, Commercial Aircraft Co., Ltd., Xi'an 710000, China; dingfangbo313@163.com
*   Correspondence: xiaosu.yi@nottingham.edu.cn; Tel.: +86-574-8818-0000 (ext. 8746)
†   These authors contributed equally to this work.

Received: 22 April 2018; Accepted: 11 June 2018; Published: 15 June 2018

Abstract: In the air and ground transportation sectors, new environmental regulations and societal concerns have triggered a search for new products and processes that complement resources and the environment. To address these issues, this article reports on current R&D efforts to develop bio-sourced materials by an international joint project. Novel bio-sourced epoxies and biocomposites were developed, characterized, modified and evaluated in terms of the mechanical property levels. Quasi-structural composite parts were finally trial-manufactured and demonstrated.

Keywords: rosin acid; itaconic acid; bio-sourced epoxy; bio-composites

1. Introduction

With the present state of composite technological development, biocomposites are understood as composites that consist of biopolymer matrices, i.e., bio-sourced resins and/or natural fiber reinforcements, e.g., plant fibers (PF) [1,2]. Besides existing applications in automobiles [3], this new member of the material family may provide an economical and environmentally-friendly alternative to glass-fiber-reinforced composites for quasi-structural applications in aircraft. The present paper provides an overview on the current development of biocomposite materials by an international joint project, ECO-COMPASS (Ecological and Multifunctional Composites for Application in Aircraft Interior and Secondary Structures, 2016–2019) [4], which is co-funded by the Chinese Ministry of Industry and Information Technology (MIIT) and the European Union, but with a special emphasis on bio-sourced epoxies and biocomposites. Quasi-structural composite parts were finally trial-manufactured and demonstrated using the epoxies as matrix resins.

2. Rosin-Sourced Epoxy as a Matrix Resin

Rosin is an abundantly available natural product, it is nontoxic and odorless, and contains various isomerized acids (>90%) and some neutral substances [5]. Reactive double bonds and carboxyl groups of rosin acids render them suitable for the Diels–Alder reaction, esterification and condensation reaction usage. Therefore, rosin acid has received increasing attention as a bio-sourced form of renewable feedstock in polymer science.

Anhydride is one of the curing agents for epoxies (EPs). Owing to their characteristic bulky hydrogenated phenanthrene ring structure, rosin acids are analogous to many aromatic compounds in rigidity. However, their curing temperature is generally high, about 200–250 °C, and the curing time is long, which limits the use of the anhydride in some products and apparatus. The complex reactions of epoxy/anhydride curing include the carboxyl group reaction with an epoxy ring in the catalyzed action of accelerator, alternate ring-opening copolymerization of epoxy, and anhydride catalyzed by amine and polyetherification at high temperatures, initiated by amine and catalyzed by hydroxyl groups [6]. These processes occur when an amine is added as the catalyst. An excellent reference [6] on this issue can be found in the literature.

In this study [7,8], an anhydride-type epoxy curing agent, maleopimaric (Scheme 1), was synthesized as a hardener from rosin acid (Scheme 1). This was supplied by the Ningbo Institute of Materials and Technology Engineering. A resin mixed with an E51-type epoxy and a solid phenolic epoxy was then prepared as the main component, together with an accelerator mixed of two amino imidazole salts as the thermally latent curing agent and also as the catalyst for the anhydride. The experimental details can also be find in reference [7]. Figure 1 exhibits the glass transition behavior of the formulated resin system. As shown, the curing degree increased with the curing temperature for a constant curing time of 3 h, as well as the glass transition temperature. The formulated matrix resin was finally designated as AGMP3600, with a bio-content of about 30%.

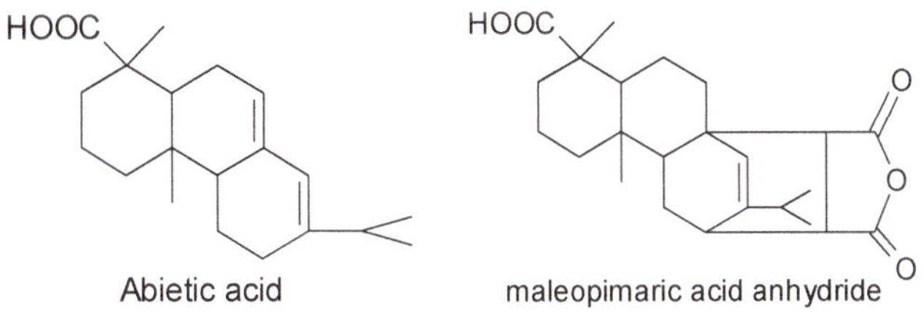

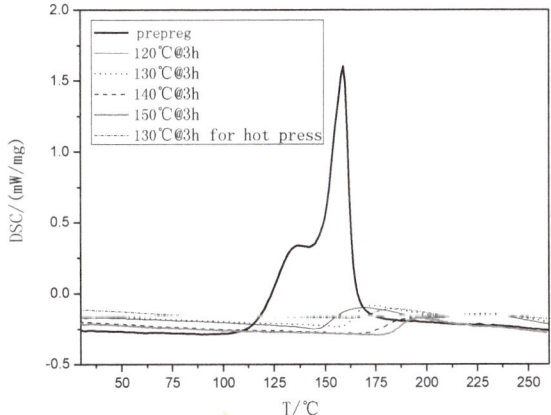

Scheme 1. Structure formula of abietic acid and maleopimaric acid anhydride.

Figure 1. Differential Scanning Calorimetry (DSC) curves of the rosin-sourced epoxy resin system designated as AGMP3600 at different temperature conditions.

Figure 2 shows the typical viscosity behavior of the trial product AGMP3600. It behaved well in the film manufacturing and subsequent prepreg production. The process condition for the prepreg using AGMP3600 as the matrix in the autoclave is shown in Figure 3.

The mechanical properties of AGMP 3600 laminates reinforced with different kinds of fibers and weaves were determined and are listed in Tables 1–4 in some cases compared with the state-of-the-art counterparts as reference.

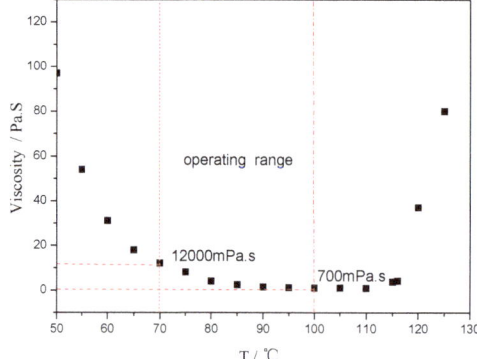

Figure 2. Viscosity vs. temperature of AGMP3600, a rosin-sourced epoxy resin system.

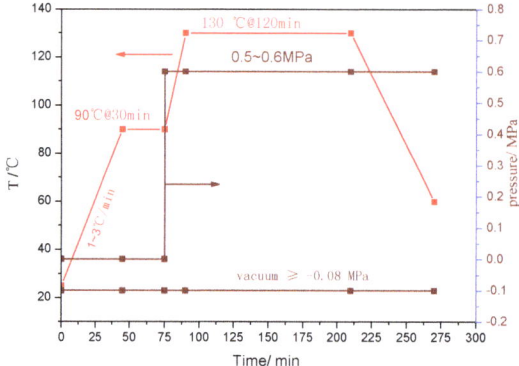

Figure 3. Process parameters for the curing of AGMP3600 prepreg in an autoclave.

Table 1. Mechanical properties of AGMP3600/EW250F (glass fabric, 8 satin, 47 vol %), a rosin-sourced epoxy laminates, and of 3233B/EW250F, a standard intermediate-temperature epoxy laminate for reference and comparison.

Properties	UNIT	3233B/EW250F	AGMP3600/EW250F
Ply thickness	mm	0.241	0.237
Bending strength	MPa	696	674
Bending modulus	GPa	19.9	21.1
Short beam shear strength	MPa	45	60.9
Tensile strength	MPa	470	540
Tensile modulus	GPa	22.5	25.3
Compression strength	MPa	474	483
Compression modulus	GPa	23.9	27

In parallel, a 180 °C/2 h cured-rosin epoxy was also developed, designated as AGMP5600, with a higher bio-content of about 40%. It is particularly interesting to note that the glass transition temperature of AGMP5600 matrix composite reinforced with EW250F glass weave was about 220 °C. Table 4 lists the mechanical properties of the laminated composite. The temperature-dependent mechanical properties were also studied, as well as those after the hot/wet exposure (1000 h/70 °C/85% r.H.)

Table 2. Mechanical properties of AGMP3600/CW3011 (carbon fiber plain weave, 200 g/m^2, 57 vol %), a rosin-epoxy laminate, and a corresponding reference laminate for comparison.

Property and Test Condition		Unit	Reference [1]	Test Result	Standard
Tensile strength warp	RT/dry	MPa	≥500	707	
Tensile modulus warp	RT/dry	GPa	65 ± 8	62.3	
Tensile strength weft	RT/dry	MPa	≥500	557	ASTM D3039
Tensile modulus weft	RT/dry	GPa	65 ± 8	60.9	
Compression strength warp	RT/dry	MPa	≥300	509	
Compression modulus Warp	RT/dry	GPa	58 ± 8	61.2	
Compression strength Weft	RT/dry	MPa	≥280	362	ASTM D6641
Compression modulus weft	RT/dry	GPa	57 ± 8	57.7	
Bending strength warp	RT/dry	MPa	≥650	883	
Bending modulus warp	RT/dry	GPa	58 ± 8	56.8	ASTM D790
Short bean shear strength	RT/dry	MPa	≥50	55.7	ASTM D2344
In plane shear strength	RT/dry	MPa	≥45	72.6	
In plane shear modulus	RT/dry	GPa	3.5 ± 1	3.84	ASTM D3518

[1] A commercial product.

Table 3. Mechanical properties of AGMP3600/A38 (carbon fiber, twill fabric, 200 g/m^2, 42 vol %), a rosin-epoxy laminate, under a hydrothermal condition of 1000 h/70 °C/85% r.H.

Properties	Unit	Humidity	Temperature/°C	Reference [1]	AGMP3600/A38	Standard
ply thickness	mm		—	0.26	0.265	
Flexural strength warp	MPa	dry	−55	—	949	
			RT	—	947	
			70	—	903	
		70 °C/wet 85%	70	—	826	ASTMD 790–03
Flexural modulus warp	GPa	dry	−55	—	50.9	
			RT	—	51.4	
			70	—	58.7	
		70 °C/wet 85%	70	—	51.8	
Interlaminar shear strength warp	MPa	dry	−55	65	62.4	ASTMD 2344/D2344M-00(2006)
			RT	60	74.8	
			70	40	65.3	
		70 °C/wet 85%	70	19	46.4	
Tensile strength warp	MPa	dry	−55	500	618	
			RT	625	667	
			70	590	714	
		70 °C/wet 85%	70	560	674	ASTMD 3039/D3039M-e1
Tensile modulus warp	GPa	dry	−55	52 ± 6	55.25	
			RT	52 ± 6	54.6	
			70	52 ± 6	53	
		70 °C/wet 85%	70	52 ± 6	49	
Poisson ratio	-	dry	RT	0.05 ± 0.005	0.052	
Compression strength warp	MPa	dry	−55	600	758	
			RT	535	651	
			70	430	630	
		70 °C/wet 85%	70	310	605	ASTMD 6641/D6641M-14
Compression modulus warp	GPa	dry	−55	46 ± 6	52.5	
			RT	46 ± 6	53	
			70	46 ± 6	52.5	
		70 °C/wet 85%	70	46 ± 6	55	
In plane shear strength	MPa	dry	−55	100	136	
			RT	95	114	
			70	80	96.8	
		70 °C/wet 85%	70	60	89	ASTMD 3518/D3518M-94
In plane shear modulus	GPa	dry	−55	4.5 ± 0.35	4.74	
			RT	3.65 ± 0.35	4.21	
			70	3.5 ± 0.35	2.63	
		70 °C/wet 85%	70	1.25 ± 0.35	2.75	

Table 3. *Cont.*

Properties	Unit	Humidity	Temperature/°C	Reference [1]	AGMP3600/A38	Standard
CAI (lay up (+/0/−/90)2 s, energy 25 J	MPa	dry	RT	180	185	ASTMD 7136/7137
Filled hole tension strength	MPa	dry	RT	180	317	ASTMD3518
Filled hole compression strength	MPa	dry	RT	250	no destroy, displacement 4 mm	ASTMD6742

[1] A commercial product.

Table 4. Mechanical properties of AGMP5600/EW250F (glass fabric, 8 satin, 47 vol %), a 180 °C cure rosin-sourced epoxy laminate, under different hydrothermal conditions.

Mechanical Properties	Unit	Test Condition	Reference [1]	AGMP5600/EW250F
ply thickness	mm	—	0.26	0.265
Flexural strength warp	MPa	−55	—	864
		RT	—	687
		70	—	650
		70 °C/wet 85%	—	431
Flexural modulus warp	GPa	−55	—	21.4
		RT	—	23.4
		70	—	22
		70 °C/wet 85%	—	22.2
Interlaminar shear strength warp	MPa	−55	88	74.6
		RT	68	59.1
		70	59	49
		70 °C/wet 85%	39.6	48.3
Tensile strength warp	MPa	−55	500	573
		RT	410	510
		70	330	460
		70 °C/wet 85%	315	—
Tensile modulus warp	GPa	−55	27 ± 6	23.9
		RT	24 ± 6	24.4
		70	24 ± 6	22.3
		70 °C/wet 85%	24 ± 6	—
Compression strength warp	MPa	−55	800	589
		RT	660	456
		70	550	397
		70 °C/wet 85%	470	384
Compression modulus warp	GPa	−55	26 ± 3	26.7
		RT	25 ± 3	25.8
		70	25 ± 3	24
		70 °C/wet 85%	25 ± 3	24.7
In plane shear strength	MPa	−55	110	113
		RT	85	90
		70	88	71
		70 °C/wet 85%	77	56.4
In plane shear modulus	GPa	−55	5.7 ± 1	4.8
		RT	4.8 ± 1	3.26
		70	3.9 ± 1	3.1
		70 °C/wet 85%	3.5 ± 1	3.13

[1] A commercial product.

3. Epoxy Resins Based on Itaconic Acid

Itaconic acid, which is also referred to as methylenesuccinic acid, is typically produced through the fermentation of carbohydrates such as glucose or starch using Aspergillus terreus. Given its strong capacity to replace petrochemicals in the chemical industry, it has been selected as one of the top 12 potential bio-based platform chemicals by the U.S. Department of Energy [9]. To the best of our knowledge, it has been widely used in the production of styrene–butadiene–acrylonitrile and acrylate latex in the paper and coating sectors.

Epoxy (EP) resin derived from itaconic acid, designated in the paper as EIA, can be synthesized following the synthetic route shown in Scheme 2. To evaluate its properties, EIA and commercial DGEBA (diglycidyl ether of bisphenol A, epoxide equivalent weight of 182–192 g/eq.) were cured with methyl hexahydrophthalic anhydride (MHHPA), respectively. The results show that EIA presented higher epoxide (0.625) and higher reactivity values than DGEBA. Relative to DGEBA, the cured EIA showed comparable or higher tensile strength (87.5 MPa), elongation at break (7.1%), flexural strength (152.4 MPa) and modulus (3430.8 MPa), and glass transition temperature (Tg = 130 °C). In addition, after co-monomers such as divinyl benzene (DVB) and acrylated epoxidized soybean oil (AESO) were introduced into the EIA/MHHPA system, dual-curing systems were formed, and the rigidity and toughness could be manipulated further via the various contents of rigid DVB or soft AESO, as shown in Figure 4, compared with data from reference published [10].

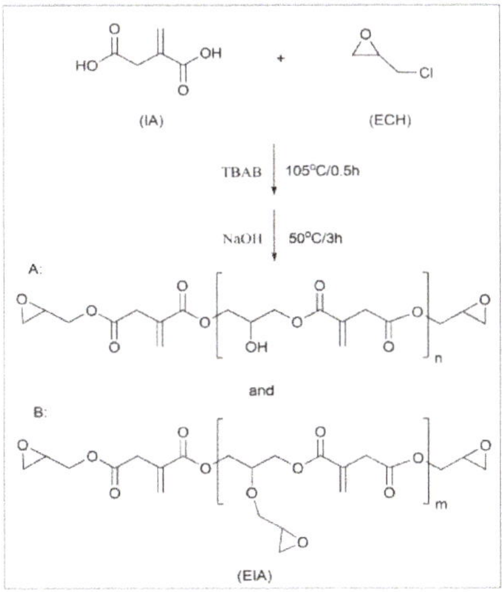

Scheme 2. Synthetic route and chemical structures of itaconic-acid-based epoxy resin (EIA).

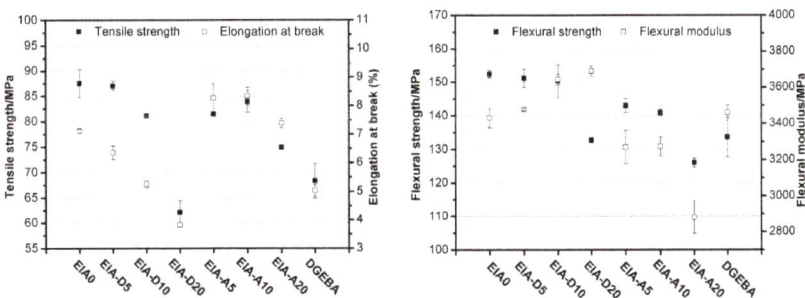

Figure 4. Mechanical properties of the cured EP resins. EIA0 and DGEBA refer to the cured samples without comonomers; D and A denote the DVB (divinyl benzene) and AESO (acrylated epoxidized soybean oil) co-monomers, respectively.

As shown in Scheme 3, direct reactions between itaconic acid and epichlorohydrin generate resin EIA, which is a mixture of different oligomers of varying molecular weights. To make the best use of the carboxyl groups and itaconic acid double bond, a trifunctional EP monomer (TEIA) was designed and synthesized (Figure 4), and it generated an extremely high epoxide value of 1.16 and a low viscosity of 0.92 Pa s at 25 °C. It is well known that low resin viscosities are beneficial for manufacturing. The lower viscosities in TEIA render it easier to process than DGEBA. In Table 5, the flexural properties of TEIA cured by different curing agents are listed. When flexible poly(propylene glycol) bis(2-aminopropyl ether) (D230) is used as the curing agent, the TEIA/D230 system shows higher flexural modulus, higher strain at break and comparable flexural strength to DGEBA/D230. When rigid MHHPA was employed as the curing agent, the flexural strength, modulus and elongation at break of the TEIA/MHHPA system exceeded those of the DGEBA/MHHPA system. These results denote that TEIA may be used as a high-performance epoxy resin [11].

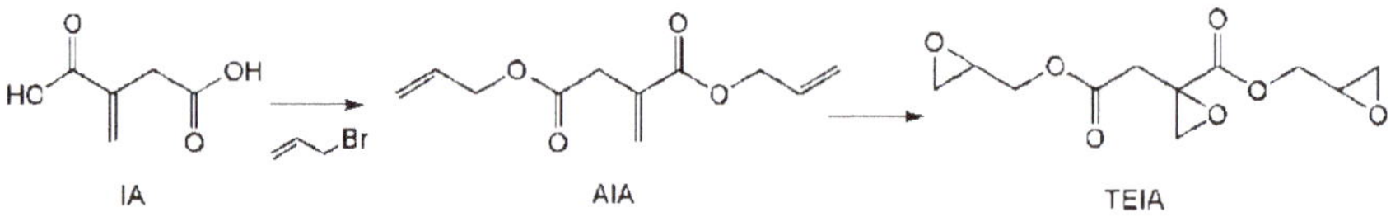

Scheme 3. Synthesis of the trifunctional epoxy resin of itaconic acid (TEIA).

Table 5. Flexural properties of cured EP resins with D230 curing agents.

Samples	Flexural Strength (MPa)	Flexural Modulus (MPa)	Elongation at Break (%)
DGEBA/D230	121 ± 1	2952 ± 18	13 ± 2
TEIA/D230	117 ± 2	3603 ± 77	21 ± 1

Given the presence of double bonds in itaconic acid and the low flame retardancy of EP resins, a flame-resistant DOPO (6H-dibenz(C,E)(1,2)oxaphosphorin-6-oxide) was chemically incorporated into the itaconic-acid-based EP resin, and a phosphorus-containing EP resin (EADI) was obtained (Scheme 4). The property study showed that the cured EADI network presents a comparable glass transition temperature and mechanical properties to those of the DGEBA system. In addition, excellent flame retardancy capacities with the UL94 V-0 grade used during vertical burning tests were observed for the EADI system. EADI may be used as a bio-based candidate for EP flame retardancy [12].

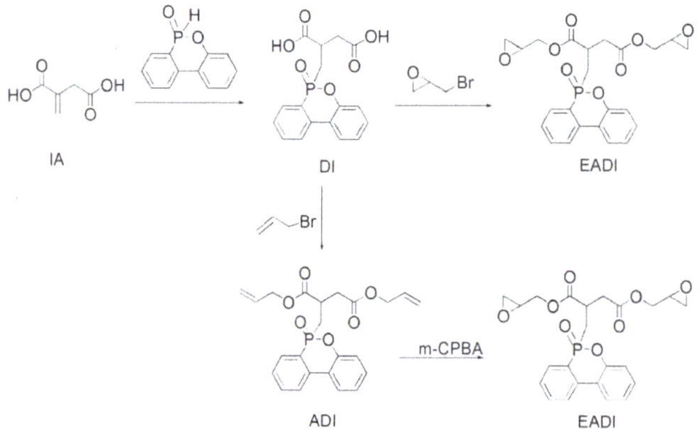

Scheme 4. Synthetic route of phosphorus-containing itaconic-acid-based epoxy resin.

4. Trial Manufacturing and Demonstration

With the promising properties, as one of the project targets, biocomposites were trial-manufactured to produce interior and quasi-structural parts for potential application in airplane and ground transportation vehicles. Figure 5 is an illustration of an interior side panel made of AGMP3600, the rosin-sourced EP, with a honeycomb sandwich core for a MA600 airplane. The side panel was manufactured using AGMP3600 prepregs in an autoclave. The composite panels are strong, lightweight, fire resistant, decorative, and impervious to mold and insects. Structure–decoration integration methods can clearly support to the production of identical or even more complex composite parts while simultaneously presenting mechanical and structural damping advantages in various applications.

Figure 5. MA600 airplane and the composite side panel made of AGMP3600/honeycomb sandwich composites (Photos courtesy of Fangbo Ding, AVIC XAC, Commercial Aircraft Co., Ltd.).

Figure 6 shows an electric race car in cooperation with Tsinghua University, China. In this case, the rosin-sourced epoxy composite was used to produce the carbon composite body with a honeycomb core. The process also used AGMP3600 prepreg. The strong conforming capacities of the materials along the curved contour were well-demonstrated. A manufacturing benefit of the biocomposites cured in an autoclave pertains to their full compatibility with the standard industrial production processes.

Figure 6. Electric race car, its body was manufactured using AGMP3600/honeycomb sandwich composites.

5. Conclusions

1. Rosin-sourced anhydride was developed and used as a hardener for epoxy to formulate a matrix resin with an imidazole-type latent catalyst for biocomposites. The mechanical properties of the biocomposites with the rosin-epoxy as matrix resins were tested, also under hydrothermal conditions. It was shown that the mechanical properties were generally comparable to the state-of-the-art, petroleum-sourced counterpart materials, but yielded a higher glass transition temperature.

2. Epoxy resin derived from itaconic acid was also synthesized. It showed comparable or higher mechanical properties and glass transition temperatures compared to a common counterpart. A phosphorus-containing epoxy was also developed by incorporating DOPO into the itaconic acid EP to formulate a flame-retardant resin system.

3. Using the rosin epoxy system, which is technologically more mature than the itaconic system, quasi-structural plant fiber reinforced components were manufactured and demonstrated for aircraft and ground transportation vehicles. The process condition was found to be fully compatible with standard industrial processes.

Author Contributions: X.Y. conceived and guided the project and study; X.Z. performed the formulation of the resins; F.D. and J.T. were responsible for the manufacturing of the structural parts.

Funding: The study was partially supported by the Chinese MIIT Special Research Program under Grant No. MJ-2015-H-G-103 and the European Union's Horizon 2020 Research and Innovation Program under Grant Agreement No. 690638. They are greatly acknowledged.

Acknowledgments: The authors are thankful to Xiaoqing Liu, Songqi Ma and Jin Zhu, NIMTE-CAS (Ningbo Institute of Materials and Technology Engineering, Chinese Academy of Sciences), for supplying the rosin-sourced polymers and related information.

Conflicts of Interest: The authors declare no conflict of interest.

References

1. Yi, X.; Li, Y. *Bio-Sourced Resins, Plant Fibers and Biocomposites*; China Construction Industry Press: Beijing, China, 2017. (In Chinese)

2. Powers, W.F. Automotive Materials in the 21st Century. *Adv. Mater. Process.* **2000**, *157*, 38–41.

3. Drzal, L.T.; Mohanty, A.K.; Misra, M. Bio-composite materials as alternatives to Petroleum-based composites for automotive applications. In *Composite Materials and Structures Center*; Michigan State University: East Lansing, MI, USA, 2000.

4. ECO-COMPASS. Available online: http://www.eco-compass.eu (accessed on 30 June 2016).

5. Coppen, J.J.W.; Hone, G.A. Gum naval stores: Turpentine and rosin from pine resin. In *NON-WOOD FOREST PRODUCTS 2*; Natural Resources Institute, Ed.; Food and Agriculture Organization of the United Nations: Rome, Italy, 1995; p. 4. ISBN 92-5-103684-5.

6. Liu, X.; Yi, X.; Zhu, J. Bio-based epoxies and composites as environmentally friendly alternative materials. In *Thermosets: Structure, Properties and Applications*, 2nd ed.; Guo, Q., Ed.; Elsevier: Amsterdam, The Netherlands, 2017; pp. 621–636. ISBN 978 0 08 101021 1.

7. Zhang, X.; Wu, Y.; Wei, J.; Tong, J.; Yi, X. Curing kinetics and mechanical properties of bio-based composite using rosin-sourced anhydrides as curing agent for hot-melt prepreg. *Sci. China Technol. Sci.* **2017**, *60*, 1318–1331. [CrossRef]

8. Lu, Y.; Zhao, Z.; Bi, L.; Chen, Y.; Wang, J.; Xu, S. Synthesis of a multifunctional hard monomer from rosin: The relationship of allyl structure in maleopimarate and UV-curing property. *Sci. Rep.* **2018**, *8*, 2399. [CrossRef] [PubMed]

9. Werpy, T.; Petersen, G. Top Value Added Chemicals from Biomass. In *Volume I—Results of Screening for Potential Candidates from Sugars and Synthesis Gas*; The National Renewable Energy Laboratory (NREL) Report DOE/GO-102004-1992; U.S. Department of Energy: Washington, DC, USA, 2004.

10. Ma, S.; Liu, X.; Jiang, Y.; Tang, Z.; Zhang, C.; Zhu, J. Bio-based epoxy resin from itaconic acid and its thermosets cured with anhydride and comonomers. *Green Chem.* **2013**, *15*, 245–254. [CrossRef]

11. Ma, S.; Liu, X.; Fan, L.; Jiang, Y.; Cao, L.; Tang, Z.; Zhu, J. Synthesis and Properties of a Bio-Based Epoxy Resin with High Epoxy Value and Low Viscosity. *ChemSusChem* **2014**, *7*, 555–562. [CrossRef] [PubMed]

12. Ma, S.; Liu, X.; Jiang, Y.; Fan, L.; Feng, J.; Zhu, J. Synthesis and properties of phosphorus-containing bio-based epoxy resin from itaconic acid. *Sci. China-Chem.* **2014**, *57*, 379–388. [CrossRef]

Review

A Review of Recent Research on Bio-Based Epoxy Systems for Engineering Applications and Potentialities in the Aviation Sector

Eric Ramon [1,*], Carmen Sguazzo [2,*] and Pedro M. G. P. Moreira [2]

1   LEITAT Technological Center, 08225 Terrassa, Barcelona, Spain
2   Laboratory of optics and experimental mechanics (LOME), INEGI Institute of Science and Innovation in Mechanical and Industrial Engineering, 4200-465 Porto, Portugal; pmoreira@inegi.up.pt
*   Correspondence: eramon@leitat.org (E.R.); csguazzo@inegi.up.pt (C.S.); Tel.: +34-937882300 (E.R.); +351-225082151 (C.S.)

Received: 1 September 2018; Accepted: 14 October 2018; Published: 16 October 2018

Abstract: Epoxy resins are one of the most widely used thermosets in different engineering fields, due to their chemical resistance and thermo-mechanical properties. Recently, bio-based thermoset resin systems have attracted significant attention given their environmental benefits related to the wide variety of available natural resources, as well as the resulting reduction in the use of petroleum feedstocks. During the last two decades, considerable improvement on the properties of bio-sourced resins has been achieved to obtain performances comparable to petroleum-based systems. This paper reviews recent advances on new bio-based epoxy resins, derived from natural oils, natural polyphenols, saccharides, natural rubber and rosin. Particular focus has been given to novel chemical formulations and resulting mechanical properties of natural derived- epoxies, curing agents or entire systems, constituting an interesting alternative for a large variety of engineering applications, including the aviation sector. The present work is within the scope of the ECO-COMPASS project, where new bio-sourced epoxy matrixes for green composites are under investigation.

Keywords: bio-based epoxy; green composite; engineering applications; aviation sector

---

1. Introduction

Thermosetting polymers are widely used within the engineering fields because of their versatility in tailoring their ultimate properties and performances in terms of strength, durability, thermal and chemical resistances as provided by the high cross-linked structure [1]. Such features make their use feasible to a broad range of applications as, for example, in bonding and adhesives for the automotive and aircraft industry [2,3], in repairing products for civil infrastructures, electronic components such as circuit boards [4], maintenance coatings for marine and multiple industries [5] or binder in laminates and composites [6]. Epoxy resins are one of the most used thermosetting systems commonly synthesized by reacting polyols, polyphenols or other active hydrogen compounds with epichlorohydrin in basic conditions [7].

According to a recent market projection [8], thermosets constitute the 14% of the overall production of polymers, mainly characterized by a wider production (82%) of thermoplastics and elastomers. Consequently, compared to thermosets, more attention has been given to thermoplastics, including bio-based ones.

Nowadays, composites based on fiber reinforced thermosetting polymer matrixes have gained importance in this field, due to their excellent mechanical properties for aircraft lightweight structures [9]. These materials are used with the aim, not to replace classic materials such as titanium and aluminum, but rather maintaining or even improving their performance. Moreover, their versatility makes them able to be used in a wide span of applications [9].

During the last fifty years, the presence of composite materials based on thermosetting polymeric matrixes for aircraft application has increased exponentially [10]. Currently, the Airbus A350 and the Boeing 787 Dreamliner are the manned passenger airplanes with more composite proportion on its structures reaching values of 50% and 53% by weight [11]. Composites used in aircrafts are usually reinforced with glass or carbon fibers, with phenolic and epoxy resins the most used thermosetting polymeric matrices for interior and secondary structures, respectively [12].

Thermosetting epoxy resins can currently encompass a wide range of properties depending on the curing agents and proportions, curing cycles and additives that can be added during their formulation [13]. This fact makes them suitable for many applications including the aerospace field. Epoxy thermosets typically used in aircrafts are based on diglycidyl ether of Bispehenol A (DGEBA). However, there are other epoxy compositions such as cycloaliphatic epoxy resins, trifunctional and tetrafunctional epoxy resins and novolac epoxy resins, although they are less used [7]. For thermosetting epoxies, the tensile strength ranges from 90 to 120 MPa with a tensile modulus ranging from 3100 to 3800 MPa [10]. Moreover, these systems usually have glass transition temperatures (T_g) that range from 150 to 220 °C, making it possible to use them as first and second aerospace resin systems [10].

Aside from the aforementioned properties, epoxy resins have two main drawbacks which are their brittleness and moisture sensitivity [10]. Besides, epoxy thermosets are difficult to be recycled and the aeronautic industry is looking for feasible alternatives to reduce the carbon footprint generated during their production [11].

Concurrently, due to an increasing demand of green industries, new natural feedstock has started being used to develop new materials. In addition, the thermosetting polymers field is currently following the same line because of the low cost, sustainability and light weight that these materials can offer.

Even though the use of bio-based epoxy systems is not yet significant in the aviation field, the aim of the present review is to highlight the possibilities of such systems, by comparing them with the corresponding petroleum-based ones and pointing out the importance to find a balance between the thermal and mechanical properties by means of chemical structure design. To that aim, in this work suitable characteristics for their application as matrix in composites for aircraft interior and secondary structures are identified within recent literature. Natural oil-based (Section 2), isosorbide-based (Section 3), furan-based epoxy systems (Section 4), phenolic and polyphenolic epoxies (Section 5), epoxidized natural rubber (Section 6), epoxy lignin derivatives (Section 7) and rosin-based resins (Section 8) are reviewed, comprising the chemical structure and the resulting mechanical properties. Glass transition temperature and viscosity are also addressed. Finally, such physical properties are summarized in Section 9, thus providing useful data that can help the future design of bio-based resins for composites in secondary and interior aviation structures. A comparison between the main mechanical properties of bio-based resins and of the petroleum-based resin systems, already used in the aeronautical field, is provided. Emphasis on potential opportunities, but also gaps and drawbacks concerning the reviewed bio-resins, are given.

2. Natural Oil-Based Epoxies

Soybean oil currently has a stable market as cooking oil, especially for the preparation of shortenings and margarine. However, soybean oil market has recently opened a new field of applications because of the capability of this oil to be epoxidized [14,15]. During the epoxidation, the double bonds of the unsaturated soybean oil are oxidated generating oxirane groups [16]. The obtained final product currently has a large market as a plasticizer and stabilizer of poly(vinyl chloride) (PVC) polymers. On the other hand, the epoxidation of vegetable oils such as epoxidized soybean oil (ESO) opens a new opportunity for the bio-based thermosetting epoxy resins. However, due to the aliphatic structure of the vegetable oils, the mechanical properties of these epoxies do not satisfy the needs for most of the aforementioned applications of thermosetting epoxy resins, and this limits their use. Nevertheless, in this section some relative recent studies of ESO thermosetting resins are described.

In a study by Zhu et al. [17], the potentialities of alternative ESO based resin systems as composite matrix were investigated, starting from the synthesis of epoxidized methyl soyate (EMS) and epoxidized allyl soyate (EAS). The two epoxidized soybean systems were added to an Epon resin, developed for pultrude composites and compared both with the neat Epon resin and a commercially available ESO resin. One-step and two-step curing processes were also analyzed. The mechanical properties of the different systems were investigated in terms of tensile and flexural behavior and are summarized in Table 1.

Table 1. Thermo-mechanical properties of the Epon/-epoxidized soybean oil (ESO); -epoxidized methyl soyate (EMS) and -epoxidized allyl soyate (EAS) systems from two-step curing adapted from [17].

Sample	T_g (°C)	Young's Modulus (MPa)	Peak Strength (MPa)	Flexural Modulus (MPa)	Flexural Strength (MPa)
Epon epoxy	74.8	3145	59	3021	110
10 wt % ESO	72.3	2807	51	3234	119
20 wt % ESO	67.0	2434	36	3090	111
30 wt % ESO	61.9	3193	60	2910	99
10 wt % EAS	75.1	2972	53	3503	127
20 wt % EAS	69.2	2979	41	3359	123
30 wt % EAS	65.0	2952	54	2979	103
10 wt % EMS	68.0	2890	45	3214	115
20 wt % EMS	63.3	2621	31	3083	110
30 wt % EMS	55.3	3145	59	2841	98

Due to the capability of the epoxidized soy ester additives of forming flexible crosslinked structures, the two synthesized soy-based resins showed better crosslinking and the resulting materials had mechanical characteristics higher than the corresponding system obtained with the commercial ESO resin. Particularly, the Epon/EAS systems provided higher values of thermo-mechanical properties.

In 2011, Altuna et al. [18] presented a study of the structure–properties relationship of blends of ESO and DGEBA cured epoxy resins. The blends were cured using a stoichiometric amount of methyltetrahydrophthalic anhydride (MTHPA) as a crosslinking agent and 1-methyl imidazole (1-MI) as a catalyst. Once cured, the thermal and mechanical properties of these samples were characterized. The obtained results indicated that increasing amounts of ESO in the blends generated a decrease in the T_g values from 108 °C (0% ESO) to 57 °C (100% ESO). The E′ in the glassy state also decreased with the content of ESO, whereas it was maintained in the rubbery phase.

The effects of ESO content were also investigated in terms of mechanical properties, in relation to impact and compression behaviors. It was observed that the compression strength and modulus remained almost constant up to 20% of ESO content and showed a decrease with higher ESO content. The impact strength of DGEBA and ESO increased about 38% in comparison with the neat DGEBA for a content of 40% ESO, whereas decreased for higher amounts of ESO. Such a behavior was ascribed to phase separation of DGEBA–40% ESO–MTHPA shown by SEM analyses.

Jin and Park [19] also investigated the effects of ESO content on the mechanical properties of DGEBA/ESO blends. Flexural properties, such as strength and elastic modulus, were derived by means of three-point bending tests, where the flexural strength had comparable values to the DGEBA neat system until a content of 40 wt % of ESO and decreased when the ESO content reached 60 wt %, and the flexural modulus showed a similar tendency. The impact strength increased with an increase of ESO wt %, while the adhesive strength increased until 40 wt % ESO content and decreased in the case of higher ESO contents.

Gupta et al. [20] investigated the influence of the DGEBA content on a epoxidized soybean oil (ESO) based systems, crosslinked by phthalic anhydride. The mechanical investigation in terms of impact, tensile and flexural testing, showed results in accordance with [18,19].

Parallel with this study, Tan et al. [21] synthesized an ESO thermally curable using 2-ethyl-4-methylimidazole (EMI) as a catalyst. They demonstrated the increment of the storage modulus as function of the EMI concentration due to the crosslinking density increment and reduction of the molecular weight. On the other hand, the T_g values (below 60.8 °C) also increased with higher EMI concentrations, but were still too low for structural and high-performance applications.

Other interesting applications of ESO as thermosetting system have been studied recently. This is the case of Cavusoglu et al. [22], that polymerized maleate half esters of oil-soluble resoles p-tertiary butyl phenol (TBP) and p-nonyl phenol (NP) with ESO. Under tensile loading conditions, the synthetized polymers showed the highest elongation for the ESO–p-NPMA–150 resin, which is in general higher than the nonpliable phenolic resins. The ESO–p-NPMA–190 resin showed the highest stress at break, and the ESO–p-TBPMA–190 showed the highest value of storage modulus at 30 °C. The mechanical properties obtained from the characterization of these samples are reported in Table 2.

Table 2. Mechanical properties of maleate half esters of oil-soluble resoles polymerized with ESO adapted from [22].

Sample	Elongation (%)	Stress at break (MPa)	Storage Modulus at 30 °C (MPa)
ESO–p-TBPMA–150 [a]	34	4	40
ESO–p-NPMA–150	128	1.5	10
ESO–p-TBPMA–190	20	12	1088
ESO–p-NPMA–190	48	13	180

[a] ESO polymerized with the maleate esters (MA) of p-TBP and p-NP.

Another application was studied by Tsujimoto et al. [23] that used ESO and polycaprolactone to generate new biodegradable shape memory polymers.

Usually, the viscosity of these systems is not optimal for application in composites, for example, during prepreg process development. However, some other applications have also been investigated for ESO, such as high temperature lubricants as indicated by Erhan et al. [15]. In that study, they determined the kinematic viscosity of ESO samples at different temperatures, which showed values of 170.87 mm^2·s^{-1} at 40 °C and 20.41 mm^2·s^{-1} at 100 °C.

On the other hand, other market established oils have also been used to generate thermosetting epoxy systems. This is the case of Linseed oil, which can be epoxidized through their multiple chain insaturations to generate epoxidized linseed oil (ELO), which is already commercial. One of the main applications of ELO is as plasticizer due to its flexible structure [24–26] also promoting thermal stabilization due to the scavenging ability of acid groups through catalytic degradation [27–29].

Researchers are now focusing on the development of thermosetting systems based on ELO, but some modifications of its chemical structure or the final composition of the materials is needed to achieve the desired properties. The study of Supanchaiyamat et al. [30] shows how ELO can be used to develop a new thermosetting resin cured with a bio-based long chain diacid (Pripol 1009) through a two steps curing process. This resin was catalyzed using different amine catalysts: Triethylamine (TEA), 1,8-diazabicycloundec-7-ene (DBU), 1-methylimidazole (1-MeIm), 2-methylimidazole (2-MeIm) and 4-dimethylaminopyridine (DMAP). Depending on the used catalyst, the viscosity of the samples ranged from 400 centipoises (cP) to 2000 cP. The obtained films showed thermal stability and excellent water resistance when 50:50 (wt %) of ELO, and Pripol 1009 and DMAP as curing agents were used. With that sample, a tensile strength of 1.65 MPa was observed.

Another study of Ding et al. [31] demonstrates how the length of the bio-derived dicarboxylic acid chains can modify the mechanical properties and the T_g of the ELO based cured epoxy resins. Using shorter dicarboxylic acids, the mechanical properties improve in terms of tensile strength, strains and modulus, toughness and T_g. However, the thermal stability decreases.

On the other hand, anhydrides have also been selected as alternative curing agents. This is the case of Pin et al. [32] that used methylhexahydrophthalic anhydride (MHHPA) and benzophenone-3,3′,4,4′-tetracarboxylic dianhydride (BTDA) as curing agents. The final resins show a bio-sourced ration over 60% and 70% for the ELO/MHHPA and ELO/BTDA respectively. These are great values compared to other green thermosetting systems currently developed. Thermal analyses reveal high thermal stability of these cured resins, with a degradation temperature that starts at 333–337 °C, which is better than some DGEBA/amine cured systems, having a degradation temperature of 300–370 °C in air. Moreover, they analyzed the viscosity changes during the curing process. The used ELO had an initial viscosity of 1200 MPa·s^{-1} which can decrease to 1 Pa·s^{-1} during the first curing stage from 25 °C to 120 °C. Such properties make these thermosets potentially appropriate in electronic applications.

Towards the development of fiber reinforced composites based on ELO, Samper et al. [29] used slate fibers as reinforcement to develop new laminate composites through resin transfer molding (RTM) process. These composite systems with ELO-based resins showed tensile strengths between 328.2 and 359.1 MPa, which are still lower values comparable with those of high-performance composites such as carbon fiber or aramid reinforced epoxies.

ESO and ELO are the most common vegetable oils able to be epoxidized due to their stability in the market and the huge amount of instaurations able to be epoxidized in their structures.

In 2012, Samper et al. [33] also investigated the production of polymers, starting from ESO, ELO and mixtures of the two oils, where phthalic anhydride (17 mol%) and maleic anhydride (83 mol %) were used as crosslinking agents and benzyl dimethyl amine (BDMA) and ethylene glycol were used respectively as the catalyst and initiator. The influence of the percentage of the two oils on the mechanical properties of the final epoxy showed that with an increase of the ESO percentage into the ESO-ELO system, the flexural strength and modulus decreased because of the smaller number of epoxy groups in ESO respect to ELO. For the same reason, the ELO-ESO resin systems also showed a decrease of the hardness when the ESO content increased. On the other hand, an increase of the impact energy was observed when the ELO percentage decreased, because of the less cross-linking which made the final epoxy system more ductile.

However, other vegetable oils have been investigated for their potential application as bio-derived thermosets. This is the case of canola oil, castor oil, karanja oil or grapeseed oil.

Espinoza Pérez et al. [34] implemented a process for the epoxidation of canola oil, obtaining sufficient conversion and scaled it up to 300 g. The process consisted of a solvent free reaction in the presence of a heterogeneous catalyst which allows for generating high content of epoxy groups on the resin. This characteristic makes the obtained resin able to be applied as matrix for composite applications. The conversion obtained was 98.5% which represented an improvement compared with other canola oil epoxidation methods [35,36].

On the other hand, Omonov et al. [37] developed a new bio-derived thermoset using epoxidized canola oil (ECO) and phthalic anhydride (PA) as curing agent. The epoxidation reaction was performed using performic acid and hydrogen peroxide. In order to prepare the curing mixture, it was necessary to form a homogeneous mixture of the ECO with the PA at higher temperatures, particularly above the melting point of PA because of its solid state at room temperature. The authors demonstrate that thermomechanical properties can be modified depending on the curing temperature and ECO/PA ratio, which makes these systems versatile for composite applications (lignocellulosic fiberboards and particleboards). They performed rheology tests to 1.0:1.0, 1.0:1.5 and 1.0:2.0 ECO/PA (mol/mol) samples at different curing temperatures, 155 °C, 170 °C, 185 °C and 200 °C. Before curing, the samples show a liquidlike behavior with optimal low viscosity that makes them optimal for their applicability. An increase in the curing temperature and PA amount decreased the gelation time of the mixtures. Conversely, the T_g also increases with an increasing amount of PA curing agent but it does not vary significantly with the curing temperature. For example, the T_g (measured with DSC) obtained for 1.0:1.0 and 1.0:2.0 ECO/PA (mol/mol) samples cured at 155 °C increased from -24.1 ± 0.3 °C to

16.1 ± 1.0 °C, but for the sample 1.0:1.0 ECO/PA (mol/mol) cured at 155 °C and 200 °C only increased from −24.1 ± 0.3 °C to −23.9 ± 0.1 which is not significant. In addition, the authors suggest that these procedures can be applied to other epoxidized natural oils by changing the reagents, curing and mixing variables.

Other natural oils, such as castor oil, have also been epoxidized with the idea of generating new bio-sourced thermosets. In this case, Park et al. [38] synthesized epoxidized castor oil using glacial acetic acid, Amberlite, toluene and hydrogen peroxide solution. Then, epoxidized castor oil was cured using N-benzylpyrazinium hexafluoroantimonate (BPH) previously tested in other studies as a cationic catalyst. The weight ratio of epoxidized castor oil/BPH was 99:1 and the curing cycles were 110 °C for 1 h, then 130 °C for 2 h and finally at 150 °C for 1 h. The glass transition temperature obtained for these systems were too low for high performance applications such as structural composites. For this reason, the authors studied blends of epoxidized castor oil with DGEBA based epoxy resins [39]. The results obtained for all the systems are shown in Table 3.

Table 3. Results of the dynamic mechanical analysis and of the diglycidyl ether of Bispehenol A (DGEBA)/epoxidized canola oil (ECO) blends cured with N-benzylpyrazinium hexafluoroantimonate (BPH) adapted from [39,40].

System (ECO [1]:DGEBA) (wt %:wt %)	T_g (°C)	Storage Modulus at 30 °C (109 Pa)	Storage Modulus at Tα + 30 °C (109 Pa)	ρ (10^{-3} mol·cm^{-3})
0:100	197	1.27	0.102	4.61
10:90	169	1.19	0.077	3.68
20:80	158	1.22	0.065	3.18
30:70	150	1.15	0.051	2.54
40:60	131	1.15	0.041	2.13
100:0	38	-	0.0079	0.57

[1] In this table ECO means "Epoxidized Castor Oil".

From these results the authors demonstrated that the glass transition temperature can be modulated with the epoxidized castor oil/DGEBA ratio with an observed decrease by increasing the epoxidized castor oil content. Regarding the mechanical behavior under flexural conditions, the strength of the ECO/DGEBA blends shows an increase with an increase of the ECO content up to a percentage of 30 wt % and does not affect the flexural modulus, whose values remain constant. Such a behavior is explained by the addition of larger soft segments of ECO into the epoxy blend, which improves its toughness.

More recently, Sudha et al. [41] also developed epoxidized castor oil/DGEBA blends at various wt % but using triethylenetetramine (TETA) as curing agent and studied the thermal and mechanical properties exhaustively. The epoxidized castor oil, analyzed in this work, had an initial viscosity of 950–1050 cP which was lower than the initial viscosity of DGEBA based epoxy. The characterization results obtained for the different blends are summarized in Table 4.

Table 4. Thermal and mechanical properties of ECO:DGEBA blends adapted from [41].

Sample (ECO [1]:DGEBA) (wt %:wt %)	T_g (°C)	Tensile Strength (MPa)	Flexural Strength (MPa)	Crosslink Density, νe ($\times 10^3$ mol/m^3)	Impact Strength Un-Notched (J/m)	Impact Strength Notched (J/m)
0:100	96.64	70.18 ± 8	95.644 ± 3	2.81	58.23 ± 6	14.05 ± 2
10:90	91.37	50.79 ± 6	83.263 ± 28	2.43	87.20 ± 4	20.27 ± 2
20:80	71.37	54.22 ± 3	100.07 ± 18	2.33	120.53 ± 11	25.33 ± 2
30:70	47.44	42.41 ± 4	81.847 ± 26	1.15	59.31 ± 1	21.80 ± 1
50:50	39.21	18.26 ± 2	40.04 ± 7	0.66	31.25 ± 3	17.25 ± 1

[1] In this table ECO means "Epoxidized Castor Oil".

As the results of this study show, the glass transition temperature and tensile strength decreases with the addition of epoxidized castor oil in the blends. The flexural strength shows comparable or higher values than the one of the pure DGEBA resin. The impact strength, measured for both Izod un-notched and notched specimens, increased with an increase of the ECO content into the blend. This is explained by the micrographs of the blends, showing the distribution of different sized particles into the cavities.

The research community has also started investigating with other less common oils such as karanja oil (KO), a natural oil extracted from karanja seeds (Pongamia Glabra) that mainly grows in India. However, not all the oil generated for human consumption is used and huge amounts of this oil finish as a waste product. With the purpose of looking for solutions for this currently unprofitable product, Kadam et al. [42] epoxidized KO using hydrogen peroxide and acetic acid. Once the natural oil was epoxidized they cured the samples using two curing agents: citric acid (CA) and tartaric acid (TA) at acid/epoxy equivalent weight ratio of 1:1. The authors tested the mechanical and thermal properties, which are summarized in Table 5.

Table 5. Thermal and mechanical properties of epoxidized karanja oil (KO) cured with citric acid (CA) and tartaric acid (TA) adapted from [42].

Sample	T_g (°C)	Tensile Strength (MPa)	Young's Modulus (E) (MPa)	Shore Hardness (A)
Bioepoxy CA [1]	112.70	10.60	2.65	56
Bioepoxy TA [2]	108.64	4.50	2.58	45

[1] Cured with citric acid. [2] Cured with tannic acid.

The differences can be seen in the glass transition temperatures being 112.70 °C for the sample cured with citric acid and 108.64 °C for the sample cured with tannic acid. Moreover, the mechanical properties depend very closely on its composition and polymerization. In the case of tensile strength, the value for the bio-epoxy cured with citric acid is higher than for the cured with tannic acid.

Other strategies to generate bio-based epoxy resins based on natural oils have been developed in recent years. This is the case of grapeseed and rapeseed oils which can be used as hardeners for epoxies. Stemmelen et al. [43], developed a novel vegetable oil-based polyamine from grapeseed oil (AGSO) through thiol-ene coupling reaction. The aminated curing agent was used to cure epoxidized linseed oil which was compared with ELO cured with tetrahydrophthalic anhydride (THPA) curing agent from the literature. The AGSO-ELO system exhibited a T_g of −38 °C, whereas the THPA-ELO system exhibited a T_g about 80 °C. The authors interpreted that this extreme change on the glass transition temperature of these two systems can be explained due to the higher molecular flexibility of AGSO compared with THPA. Manthey et al. [44], explored the thermo-mechanical properties of epoxidized hemp oil based samples (EHO), as possible matrix material for jute fibre-reinforced bio-composites. A maximum concentration of 30% of EHO in a synthetic bisphenol A diglycidyl ether-based epoxy control, R246TX cured with a blend of triethylenetetramine and isophorone diamine, was investigated. Such an epoxy system was also studied in comparison with an ESO-based system with the same bio content. The EHO-based epoxy blends displayed slightly higher mechanical properties than the corresponding ESO-based system. Furthermore, the EHO-based resin blend, when used as matrix for the jute fibre-reinforced biocomposites, was found to be competitive with the commercially produced ESO-based system.

In the work of Akesson et al. [45], two ESO based resins were synthesized and cured by means of ultraviolet (UV) irradiation: particularly one resin was a synthesized methacrylic anhydride modified soybean oil (MMSO) type and another resin was an acrylated epoxidized soybean oil (AESO). The MMSO resin had a storage modulus of 1800 MPa and a glass transition temperature above 150 °C and showed a tensile strength at break, over five specimens, of 24.4 ± 3.6 MPa.

3. Isosorbide Based Epoxy Resins

Isosorbide is a bio-derived diol with a structure based on two fused furan rings. It is obtained from starch originated from depolymerization of biomass, which is firstly hydrolyzed to generate D-glucose [46]. Then, the D-glucose is hydrogenated to generate sorbitol which is dehydrated to obtain the isosorbide in a multi-step process [46]. Due to the versatility of its hydroxyl groups, isosorbide can be easily derivatized for use in some applications for pharma, detergency, cosmetics, or as a stabilizer or plasticizer [47]. Moreover, isosorbide has recently become a solution for the generation of new bio-based thermoplastics and thermosets, such as epoxy resins. As reported by Hong et al. [48], isosorbide rigidity and thermal stability provided by its structure allows its use as an alternative monomer to bisphenol-A (BPA) in epoxy resins: diglycidyl ether of isosorbide (DGEI). This monomer can be obtained through different ways: epoxidation of isosorbide diallyl ether, reaction of isosorbide with epichlorohydrin in presence of Lewis acid or alkali hydroxides, or through reaction of isosorbide disodium alcoholate with epichlorohydrin. In that study, isosorbide resins and control DGEBA resin were cured with diethylene triamine (DETA) and isosorbide diamine (ISODA). An extensive experimental campaign was also conducted to determine the thermo-mechanical properties of the isosorbide-based resin from corn and some of the results are reported in Table 6.

Table 6. Thermal and mechanical properties of isosorbide-based resin adapted from [48].

Sample	T_g (°C) *	Tensile Strength (MPa) **	Young Modulus (MPa) **	Flexural Modulus (MPa) **	Impact Strength (J/m) **
DGEBA/DETA	129 (134)	26 (8.2%)	1389 (4.7%)	3061 (1.6%)	60 (1.2%)
DGEBA/ISODA	74 (79)	67 (4%)	1825 (4.5%)	3364	94 (65%)
DGEI(mono)/DETA	75 (76)	62 (9%)	1798 (1.2%)	4027	72 (16.8%)
DGEI(mono)/ISODA	32 (43)	41 (21%)	1532 (2.6%)	1168	65 (23%)
DGEI(polymeric)/ISODA	36 (43)	52 (8.1%)	2461 (9.5%)	3520	57 (14.7%)
DGEI(polymeric)/DETA	48 (63)	52 (18%)	1774 (8%)	2747	113 (33%)

* Values in brackets are obtained after post-curing at 150 °C for 2 h. ** Values in brackets-coefficient of variation. Those without coefficient of variation are based on one specimen.

The isosorbide-based resins showed glass transition temperatures of about 60 °C lower than those from DGEBA/DETA, and comparable or better tensile and impact strength than commercial epoxy resins. Moreover, rheological tests were performed to the curing mixtures. The initial viscosity values for the DGEI/ISODA and DGEI/DETA were found to be below the 10,000 cP, which are values similar to the commercial DGEBA (from 10,000 to 25,000 cP using the same curing agents), as compared with natural oils.

Feng et al. [49], prepared bisisorbide diglycidyl ether through isosorbide reaction with epichlorohydrin under alkali conditions. The resulting resin was then cured using an aliphatic amine Jeffarnine T403 and compared with EPON 826, a diglycidyl ether of bisphenol A cured using the same curing agent. The tensile strength of the cured isosorbide-based epoxy and the DGEBA were on average 68.8 MPa and 66.2 MPa, respectively. This means that the tensile strength of the bisphenol A is 96% of the isosorbide-based epoxy. The impact strength of the isosorbide-based epoxy was 40% higher compared with the DGEBA based epoxy with average values of 3.87 J/cm and 6.42 J/cm respectively. However, the T_g of isosorbide-based epoxy systems was usually lower than the T_g of the DGEBA based epoxy thermosets (48 °C compared with 90 °C respectively). This is because of the high hydrophilicity of the isosorbide and the diamine, but it can be optimized with other curing agents such as methyl-5-norbornene-2,3-dicarboxylic anhydride with catalyst benzyl dimethyl amine (BDMA) or with 4,4'-(hexafluoro-isopropylidene) diphthalic anhydride that allows an increase of its T_g to 113 °C and 200 °C respectively.

Łukaszczyk et al. [50], compared the properties of isosorbide epoxy resins (IS-EPO) cured with phthalic anhydride (PHA), tetrahydrophthalic anhydride (TPHA), triethylenetetramine (TETA), isophoronediamine (IPHA) with the properties of Epidian 5 DGEBA based epoxy resin. This study also confirmed that, in most of the cases, the mechanical properties of isosorbide based epoxy resins, such as flexural and compression strength and modulus, Brinell hardness, Izod impact strength, were found to be comparable to, or better than, the properties of DGEBA based epoxy resins, as reported in Table 7.

Table 7. Mechanical and T_g properties of Epidian 5 and isosorbide based epoxy resin systems cured with phthalic anhydride (PHA), tetrahydrophthalic anhydride (TPHA), triethylenetetramine (TETA) and isophoronediamine (IPHA) adapted from [50].

Sample Composition	T_g (°C)	Flexural Strength (MPa)	Compression Strength (MPa)	Brinell Hardness (MPa)	Izod Impact Strength (kJ/m^2)
Epidian 5/PHA	171	158.4	290.8	198.0	7.2
IS-EPO/PHA	108	225.5	254.1	202.4	30.9
Epidian 5/THPHA	172	27.9	122.2	209.4	4.1
IS-EPO/THPHA	95	100.5	88.8	214.3	2.9
Epidian 5/TETA	116	170.8	234.2	212.1	9.5
IS-EPO/TETA	49	228.3	311.6	193.8	20.8
Epidian 5/IPHA	141	175.4	193.9	231.2	13.5
IS-EPO/IPHA	73	158.5	318.1	205.7	33.8

However, the T_g of isosorbide-based epoxy resins is, in all cases, lower than the DGEBA based ones. The kinematic viscosity obtained for the IS-EPO samples before the curing was 60,120 cP, which is higher than the values of the Epidian based samples (25,000 cP).

In 2013, Sadler et al. [51] synthetized isosorbide-methacrylate (IM) by the direct esterification of isosorbide using methacryloyl chloride or methacrylic anhydride and a base catalyst. The IM was also blend in a vinyl-ester (VE) resin. The IM-based VE resin possessed a T_g higher than that of any known commercial vinyl ester resins. Very interesting mechanical properties were found out, such as quite high flexural strength and modulus of the neat IM. Those properties make it classifiable in the range of high-performance materials.

In 2017, Liu et al. [52] synthetized isosorbide-methacrylate (IM), from isosorbide with methacrylate anhydride (MAA) via a solvent-free, ultrasonic-assisted method. It was then to copolymerize an acrylated epoxidized soybean oil (AESO), finally obtaining a biobased thermosetting resin (IM-AESO). A second system was obtained by further modifying the AESO with MAA to replace the hydroxyl groups with methacrylate groups, thus generating a resin (IM-MAESO). The soybean oil-based resins blended with IM as an RD shows superior processability due to the low viscosity. The mechanical properties, such as flexural strengths, flexural moduli, storage moduli were investigated, showing high values. These aspects together with the superior processability make these two resins systems suitable for potential application in fiber-reinforced composites.

4. Furan Based Epoxy Resins

Furan resins have recently appeared as a new bio-based alternative to phenolic resins due to their aromaticity and mechanical properties. The origin of furanyl building blocks usually starts from furfural, which is transformed to furfuryl alcohol through their hydrogenation [53,54]. Furfural is obtained by chemical dehydration of five carbon carbohydrates, such as xylose and arabinose by fractionation of hemicellulose from bagasse, left over from sugar cane processing and also from corn cobs or other biomass waste [55,56]. The structure of furfural is shown in Figure 1.

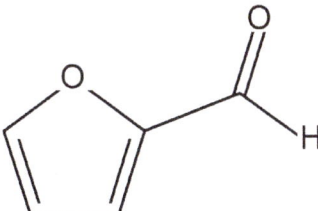

Figure 1. Furfural structure.

Apart from furfural and furfuryl alcohol derived from five carbon carbohydrates, there are also other building blocks able to be polymerized to obtain different furan resins with some promising properties that could be useful for aircraft applications. However, these are still under development and they are not yet implemented in the industrial field. One important example is 5-Hydroxymethylfurfural HMF, which is obtained by hexoses dehydration [53].

Main monomers used for furan building blocks can be obtained from bio-sources which allow reduction of carbon footprint of final polymeric material. A list of the most important ones and its market availability is shown in Table 8.

Table 8. Bio-sourced building blocks for furan resins.

Furanyl Monomer	Source	Status
Furfural (F)	Sugar cane bagasse or corn cobs (derived from pentoses)	Commercial
Furfuryl alcohol (FA)	from furfural.	Commercial
5-Hydroxymethylfurfural (HMF)	Plant based sugars (derived from hexoses)	Commercial
2-Furfurylmethacrylate (FM)	From HMF or furfural	Non-commercial
Bis-2,5-hydroxymethylfuran (BHMF)	From HMF or furfural	Non-commercial
2,5-Furandicarboxylic acid (FDCA)	From HMF or furfural	Commercial

Currently, epoxidized furanyl building blocks are not up-scaled and commercially available. However, some researchers have studied the functionalization of raw furan monomers F and HMF in order to synthesize new epoxy building blocks. This is the case of Cho et al. [57], who reported the preparation of furan monoepoxide, 2,5-bis[(2-oxiranyl-methoxy) methyl]-furan (BOF) and bis-furan diepoxide. These monomers were photo-cured using IRGACURE 250 and triphenylsulfonium hexafluoroantimonate salt (TSAS, 50% in propylene carbonate) as cationic crosslinking agents and their properties for adhesive industry were investigated.

The furanic compounds functionalized by epoxide group(s) were used for bonding polycarbonate (PC) by cationic photo-curing and with petrochemical-based phenyl glycidyl ether (PGE) having a phenyl ring. Specimens of the bonded joint for the measurement of tensile-shear strength were prepared and tested. PC joints bonded by both compounds biomass-based furan monoepoxide (FmE) and PGE petroleum-based epoxy were prepared. The strength of bonded PC joints (FmE) showed better performance as an adhesive than PGE. The authors attributed the lower tensile-shear strengths of PC joints bonded by Furan diepoxide (FdE) and bis-furan di-epoxide (bFdE) to their rigid structure and hydrophobic property.

Hu et al. [58] synthesized 2,5-Bis[(2-oxiranylmethoxy)methyl]-furan (BOF) and 1,4-Bis[(2-oxiranylmethoxy)methyl]-benzene (BOB) monomers and cured them using PACM and EPIKURE W as curing agents. Both monomers were low viscosity liquids with really good processability. The authors also blended the two furanyl monomers with DGEBA and performed a thermo-mechanical characterization. The results of the T_g values (obtained from DMA analysis) are shown in Table 9.

From these results, the authors determined that for the samples cured with EPIKURE W the T_g values are always higher compared with the samples cured with PACM. This fact is due to the aromatic structure of EPIKURE W. Another observation was that for BOF systems cured with either curing agents the T_g values and the storage modulus are higher than for the BOB systems. A possible explanation for this success could be the higher possibility of creating hydrogen bonds between the oxygen atoms of the furan rings and the hydroxyl groups created during the epoxy group opening during the reaction with the curing agents for BOF. Another explanation could be the differences in rotation of the furan rings due to their structures.

In another study, Hu et al. [59] also developed a new bio-based curing agent based on furan chemistry in order to create full furan based thermosetting resins. To achieve this aim, the authors prepared 5,5'-Methylenedifurfurylamine (DFDA) and 5,5'-Ethylidenedifurfurylamine (CH$_3$-DFDA) as curing agents and compared the T_g values obtained from DMA thermograms. BOF/DFDA and BOF/CH$_3$-DFDA samples exhibited T_g values of 69 and 62 °C, respectively, which allows for explaining the positive effect of the methyl group on the T_g value for CH$_3$ DFDA cured BOF. On the other hand,

PACM cured BOF showed a T_g of 72 °C, a higher value compared with DFDA and CH3- DFDA cured samples due to the connection to the furan rings through the methylene groups and amine groups. The samples of DGEBA cured with the same curing agents showed higher T_g values as it was expected (121 °C, 128 °C and 168 °C for DGEBA/DFDA, DGEBA/CH$_3$-DGBA and DGBA/PACM respectively).

Table 9. T_g values of 2,5-bis[(2-oxiranyl-methoxy) methyl]-furan (BOF), 1,4-Bis[(2-oxiranylmethoxy) methyl]-benzene (BOB) and DGEBA mixtures cured with PACM and EPIKURE W adapted from [58].

Weight Ratio Monomers (BOF:BOB:DGEBA)	T_g (°C)	
	PACM	EPIKURE W
100:0:0	71 (80) [1]	88 (94)
70:0:30	96 (106)	114 (120)
50:0:50	111 (121)	133 (139)
30:0:70	131 (140)	153 (160)
0:100:0	55 (63)	80 (90)
0:70:30	84 (94)	104 (103)
0:50:50	103 (114)	126 (136)
0:30:70	124 (134)	148 (159)
0:0:100	167 (176)	185 (198)

[1] T_g in parentheses is the value obtained from maximum of tan δ curve from DMA.

The authors investigated the tensile and fracture properties of BOF/PACM, BOB/PACM and DGEBA/PACM samples. In terms of tensile maximum stress, the BOF/PACM, BOB/PACM shows slightly lower values when compared to the DGEBA/PACM system; while in terms of the critical stress intensity factor (K_{1C}) and critical strain energy release rate (G_{1C}) values, BOF/PACM sample shows superior fracture toughness properties compared with BOB/PACM and DGEBA/PACM samples. Further studies about the mechanical characterization are also reported in Hu et al. [60].

On the other hand, Deng et al. [61] prepared diglycidyl ester of 2,5-furandicarboxylic acid (DGF) and compared its properties to the petroleum-based counterpart diglycidyl ester of terephthalic acid (DGT). The curing agents used were methylhexahydrophthalic anhydride (MHHPA) and poly(propylene glycol)bis(2-aminopropyl ether) (D230) as the rigid and soft curing agents respectively. The synthesis of DGF was performed using 2,5-Furandicarboxylic acid (FDCA) as raw material which was first used to generate bis(prop-2-enyl)furan-2,5-dicarboxylate (FDCE). The equivalent ratio of MHHPA and DGF or DGT was 1:1 respectively and the mixtures were cured at 100 °C during 2 h, 130 °C during 2 h and 150 °C during 2 h. The thermal and mechanical properties of the different samples developed in this study are listed in Table 10.

Table 10. Thermal and mechanical properties of diglycidyl ester of 2,5-furandicarboxylic acid (DGF), diglycidyl ester of terephthalic acid (DGT) and DGEPA resins cured with methylhexahydrophthalic anhydride (MHHPA) and D230 adapted from [61].

Sample	T_g (°C)	Tensile Strength (MPa)	Tensile Modulus (MPa)	Flexural Strength (MPa)	Flexural Modulus (MPa)
DGF/MHHPA	152	84 ± 4	3000 ± 50	96 ± 3	3100 ± 110
DGT/MHHPA	128.8	78 ± 2	3080 ± 80	90 ± 5	2950 ± 40
DGF/D230	101.2	68 ± 3	2700 ± 110	75 ± 2	2500 ± 90
DGT/D230	91.8	64 ± 2	2800 ± 60	73 ± 3	2400 ± 100
DGEBA/MHHPA	125	68	2900	135	3400
DGEBA/D230	97	NA	NA	121	2950

The results obtained from the thermal and mechanical analysis show that DGF systems have higher T_g, and mechanical properties compared with DGT systems. Moreover, these two systems show comparable properties to the DGEBA based ones. This fact allows for confirming that bio-based epoxy resins obtained from FDCE have a high potential as petrol-sourced epoxy alternatives.

5. Phenolic and Polyphenolic Epoxies

Polyphenols can be classified in different classes depending on the strength of the phenolic ring: phenolic acids, flavonoids, stiblins, phenolic alcohols and lignans [62]. Natural polyphenolic and phenolic structures are interesting due to their mechanical properties and thermal stability that they confer to the epoxy resins. For this reason, some researchers have focused their activity on the functionalization and preparation of epoxy resins based on natural phenolic or polyphenlic structures.

Tannins are natural polyphenolic structures generally obtained from black mimosa bark, quebracho wood, oak bark, chestnut wood, mangrove wood, Acacia catechu, Uncaria gambir, sumach, myrabolans, divi-divi, algarobilla chilena, tara, and bark of several pine species [63]. Tannins can be mainly classified depending on their structures as hydrolysable tannins and polyflavonoid tannins [63].

Catechin is one of the main components of tannins and one of the most studied natural polyphenol for the generation of new epoxy resins. Epoxidation of catechin was firstly examined by Nouailhas et al. [64] as a laboratory model with the aim of studying the reactivity of tannins as potential substitutes for BPA. The authors also used resorcinol and 4-methylcatechol in order to mimic the A- and B-rings of catechin. The glycidylation of rescorcinol, 4-methylcatechol and catechin was performed using epichlorohydrin. The final products were then identified, and their structures were confirmed by Nuclear Magentic Resonance (NMR) and Furier-transform Infrared Spectroscopy (FTIR). Finally, some formulations were prepared blending glycidyl ether of catechin (GEC) with DGEBA resin in order to reduce the initial viscosity of initially solid GEC. Then the samples were analysed by Dynamic Mechanical Analysis once cured at 60 °C for 24 h with Epamine PC 19. This curing agent provides low viscosity and fast curing capabilities. The results obtained are shown in Table 11.

Table 11. T_g values and mechanical properties of DGEBA/glycidyl ether of catechin (GEC) blends cured with Epamine PC 19 adapted from [64].

Sample	T_g (°C)	Storage Modulus (GPa) at 30 °C	Storage Modulus (GPa) at T_g + 30 °C
DGEBA	209	2.81	0.019
75DGEBA/25GEC	221	2.46	0.016
50DGEBA/50GEC	202	2.40	0.014

Some of the authors of the previous work also participated in a study published in 2014 [65] with the aim of continuing the study on model phenolic monomers but now using tannins directly extracted from green tea. In this work, Benyahya et al. [65] also used epichlorohydrin for the epoxidation of the tannins extracted from green tea with BnEt3NCl as phase transfer catalyst. After curing blends of GEC, glycidyl ether of green tea extract (GEGTE) and a DGEBA epoxy resin (DER 352) with isophorone diamine (IPDA) in a 1:1 molar ratio of epoxy groups, dynamic mechanical analysis was also performed. Some of the main results are shown in Table 12.

Table 12. T_g values and mechanical properties of GEC, glycidyl ether of green tea extract (GEGTE) and DGEBA based epoxy resins cured with isophorone diamine (IPDA) adapted from [65].

Sample	T_g (°C)	Storage Modulus (GPa) at 30 °C	Storage Modulus (GPa) at T_g + 30 °C
GEGTE-IPDA	142	2.34	0.0593
GEC-IPDA	179	1.50	0.0364
DER352-IPDA	140	1.29	0.0136

The authors observed similar values for GEGTE-IPD and DER352-IPD systems, but a higher value was observed for the GEC-IPD. This could be explained by the higher presence of glycidyl groups that allows higher crosslinking density. Another explanation that the authors give is that, although GEGTE-IPD could have high functionality, a reduction of the crosslinking density could be caused by the presence of dimmers of gallate. However, gallic acid groups could reinforce the GEGTE-IPD system that shows higher storage modulus than GEC-IPD and DER352-IPD.

Basnet et al. [66] used a lignin derivative as curing agent. The glycidylation process of catechin was performed using epichlorohydrin under alkaline conditions and with the presence of TMAC as phase transfer catalyst. The authors also synthesized glycidyl ether of heat dried green tea extract (GEHDGTE), glycidyl ether of freeze-dried green tea extract (GEFDGTE), glycidyl ether of standard catechin (GEC), as well as EP828 (a commercial DGEBA based epoxy resin) and compared their properties. Glass transition temperature and flexural strength results are shown in Table 13.

Table 13. T_g values and flexural strength of GEC, glycidyl ether of heat dried green tea extract (GEHDGTE), glycidyl ether of freeze-dried green tea extract (GEFHDGTE) and BPA based epoxies cured with a lignin derivative adapted from [66].

Sample	T_g (°C)	Flexural Strength (MPa)
GEC-Lignin	178	63
GEHDGTE-Lignin	155	56
GEFHDGTE-Lignin	173	40
BPA-Lignin	150	29

In this study, the authors demonstrated the potential applicability of these materials for electronic applications also due to the temperature resistance of the formulations that were similar or higher than for DGEBA based epoxy resins. The mechanical flexural properties were found to be much higher than the corresponding DGEBA-Lg. The authors attributed the higher rigidity of the new lignin-based compounds to the combination of aromatic raw material of low molecular weight and the hard segment of the lignin, as curing agent, which has average molecular weight.

Other phenolic biomolecules, such as gallic acid, have also been considered as a potential source for bio-based epoxy thermosetting resins. Gallic acid or its derivatives are present in grapes, tea, gallnuts, oak bark, some fruits, vegetables, honey and can also be found in plant tissues [67–71]. The phenolic structure of gallic acid is shown in Figure 2.

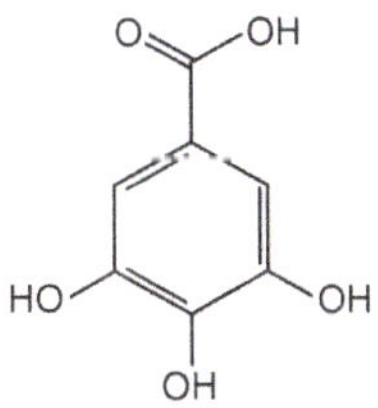

Figure 2. Gallic Acid.

Tomita et al. [72], presented a patent where they epoxidized gallic acid or tannic acid with epichlorohydrin in presence of a phase transfer catalyst. They cured the resulting epoxies using 3,6-endomethylene-1,2,3,6-tetrahydromethylphthalic anhydride (Methylhimic Anhydride) and benzyldimethylamine at different proportions. The curing conditions were 100 °C for 3 h and then 150 °C for 15 h. Finally, they compared the heat deflection temperature of these systems with Epikote 828, a DGEBA derivative epoxy resin and the results showed higher values for the bio-based systems.

On the other hand, Aouf et al. [73] synthesized epoxy resin prepolymers based on natural phenolic compounds such as gallic and vanillic acids. The strategy used by those authors was to synthesize the allylated phenolic before the epoxidation of the double bonds using caprylic acid as oxygen carrier and Novozym 435, an immobilized lipase B from Candida Antarctica, as a biocatalyst. This method was then compared with the epoxidation using the peracid mCPBA, which showed lower yields compared with the curing process using mCPBA.

Recently, Tarzia et al. [74], slightly modified the epoxidation method previously used by Aouf et al. [73]. In this study, tri- and tetra-glycidyl ethers of gallic acid (GEGAs) were obtained and cured with three different curing agents: a cycloaliphatic primary amine, isophorone diamine (IPDA), and Jeffamine D-230 (DPG). Moreover, they used *N,N*-dimethylbenzylamine (BDMA) as an ionic initiator. The T_g values and mechanical properties of each system are collected in Table 14.

Table 14. T_g values and mechanical properties of GEGA and DGEBA cured with IPDA, DPG and *N*-dimethylbenzylamine (BDMA) adapted from [74].

Sample	T_g (°C)	Tensile Modulus (GPa)	Tensile Strength (MPa)	Elongation at Break (%)
GEGA/IPDA	158	3.6 ± 0.3	43.1 ± 13.1	1.4 ± 0.3
GEGA/DPG	98	3.5 ± 0.2	70.6 ± 2.9	6.1 ± 0.6
GEGA/BDMA	136	3.2 ± 0.2	31.2 ± 2.3	1.1 ± 0.1
DGEBA/IPDA	-	3.1 ± 0.8	34.1 ± 2.0	1.7 ± 0.2
DGEBA/DPG	-	3.0 ± 0.2	116.4 ± 7.0	8.6 ± 0.3

The GEGA/DPG system is the one with lower T_g values which can be explained due to the flexibility and linearity of the DPG backbone. On the other hand, IPDA showed the higher T_g value. This can be explained due to a higher crosslinking density and more rigidity due to the IPDA ring. Comparing the mechanical properties, DPG cured system showed improved strength compared with the IPDA and BDMA systems. However, DGEBA/DPG system showed higher tensile strength and elongation at break compared with the GEGA/DPG system. This result can be attributed to the GEGA structure, which is stiffer due to the single aromatic ring compared with DGEBA and with higher functionality. Regarding the elastic modulus, the GEGA systems showed an elastic modulus slightly higher than the one corresponding to the DGEBA systems. The viscosity value obtained for the GEGA precursor was found to be 2000 cP before curing. This value is lower than the one of typical commercial DGEBA systems, that ranges from 11,000 to 15,000 cP [75] and can reach higher values depending on the formulation.

In 2013, Cao [76] derived a bio-based epoxy monomer GA-II from renewable gallic acid. An extensive experimental campaign was performed by investigating the mechanical properties in terms of tensile strength and modulus, where values of 66 MPa and 1970 MPa were respectively reached, and flexural strength and modulus, where values of 128 MPa and 3050 MPa were reached.

Tannic acid has also been considered as an interesting candidate to be used as a curing agent for epoxy resins. Tannic acid is a polyphenol currently commercial as a mixture of gallotannins. Some studies have been carried out with the aim of generating epoxy thermosets from tannic acid. This is the case of Shibata et al. [77], that used commercial tannic acid as curing agent to prepare composites with glycerol polyglycidyl ether (GPE) and sorbitol polyglycidyl ether (SPE). The mechanical properties and T_g values of each system are shown in Table 15.

Table 15. T_g values and tensile strength of glycerol polyglycidyl ether (GPE) and sorbitol polyglycidyl ether (SPE) cured with TA adapted from [77].

Sample	Epoxy/–OH	T_g (°C) [TMA]	Tensile Strength (MPa)
GPE-TA	1.0	87.3	36.7
SPE-TA	1.0	106.6	60.6

SPE cured with TA samples showed higher T_g values compared with GPE samples. This result can be explained by the higher functionality of SPE compared with GPE. For the same reason, the mechanical properties of SPE have a higher result than the GPE ones.

Cardanol is another phenolic molecule obtained from cashew nut shell liquid extracts [78,79]. The structure of cardanol is shown in Figure 3.

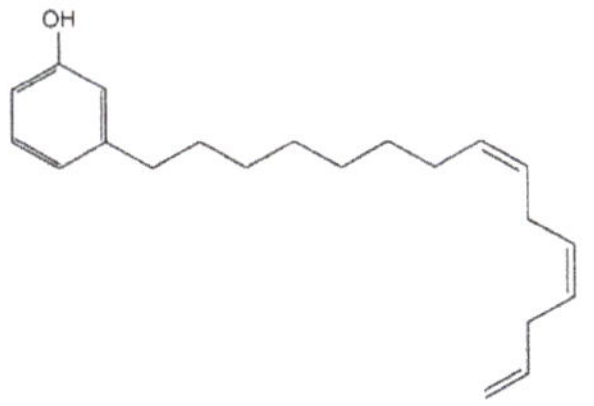

Figure 3. Cardanol.

One of the multiple and recent proposes for cardanol is to use it as a green raw material for the production of various polymers types, such as polyurethanes or Novolac resins. Green epoxy resins have been produced in some studies through the epoxidation of cardanol or using it as a curing agent. In 2004, Maffezzoli et al. [80] demonstrated the effectiveness of cardanol as a green building block for thermosetting epoxy matrix. The samples were cured using an amine curing agent and an acid catalyst and mixed with DGEBA resins. The sample with the resol and DGEBA in a stoichiometric ratio and with acid catalyst showed a tensile strength of 12 ± 2.2 MPa and a modulus of 864 ± 79 MPa; its values of deformation under tensile load also resulted to be higher than the resol DGEBA system not containing acid catalyst.

In 2008, Unnikrishnan et al. [75], epoxidized cardanol using epichlorohydrin in presence of caustic soda catalyst and compared the results with DGEBA based resins. Some of the mechanical properties obtained from these samples have been collected in Table 16.

Table 16. Mechanical properties of cured BPA/cardanol epoxy blends adapted from [75].

Sample	Tensile Strength (MPa)	Compressive Strength (MPa)	Elongation at Break (%)	Flexural Strength (MPa)	Impact Strength (Izod, J/m)	Young's Modulus (MPa)
Commercial DGEBA	48.0	108	3.1	91.45	28.5	2420
BPA/cardanol epoxy (80:20) [a]	31.7	92.55	5.68	80.8	22.25	2045
BPA/cardanol epoxy (50:50)	23.5	70	8.12	71.45	20.4	1926

[a] Molar compositions.

The authors of this study demonstrated that the introduction of epoxidized cardanol in DGEBA resins produces a decrease of the tensile, compressive and Izod impact strengths and it does not have much influence on the elongation-at-break. However, the introduction of cardanol can reduce the brittleness of the epoxy systems making them more flexible, as shown by the higher values of flexural modulus and strength of these values, comparable with the commercial liquid epoxy resins which have a viscosity of 11,000–15,000 cP [75].

Jaillet et al. [81], used a commercial epoxidized cardanol to generate new green epoxies in 2014. The commercial name of the epoxidized cardanol is NC-514 from Cardolite which was then cured using IPDA and Jeffamine D400. The maximum T_g values obtained for each system were 50 °C when cured with IPDA and 15 °C when cured with Jeffamine D400. These values are too low for most epoxy applications that require higher thermal and mechanical properties. To solve this problem, Darroman et al. [82] proposed to use epoxy blends with epoxidized cardanol to improve the final thermomechanical properties. The authors used isophorone diamine (IPDA) and Jeffamine T403 as amine curing agents and blended the epoxidized cardanol with epoxy sorbitol and epoxy isosorbide. The results showed that the T_g value increases when the epoxy sorbitol and epoxy isosorbide content increase, obtaining maximum values below 82 °C and 109 °C respectively, when cured with IPDA.

The most recent study was carried out by Atta et al. [83], that prepared cardanol novolac epoxy (CNE) resins by reacting cardanol with formaldehyde, followed by epoxidation in glacial acetic acid and epicholohydrin. Then, they prepared a cardanol polyamine hardener (CPA) to cure the cardanol novolac epoxy resins. The final cured systems showed T_g values from 50 to 84 °C and its low viscosity values of 1150 and 2800 cP for CNE and CPA respectively, makes them suitable for marine coating applications.

6. Epoxidized Natural Rubber

Natural rubber is obtained from the Brazilian rubber tree and it is one of the most used polymers produced by plants for more than 40,000 products [84]. This natural feedstock can be epoxidized with peracids by the double bonds that are present on its structure [85]. Epoxidized natural rubber is currently a commercial product used for various applications such as toughener [86], compatibilizer [87], adhesive industry [88] and blends an reinforcements [89,90]. Epoxidized natural rubber is commercially available under the name of ENR-25, ENR-50 or ENR-75 depending on the epoxy content [91]. In 2013, Hsmzah et al. [92] elucidated the structure of the ENR-50 by HNMR and studied their modification using a cyclic dithiocarbonate.

In other studies, the authors used ENRs as modifiers. This is the case of Mathew et al. [93], that used epoxidized natural rubbers (ENRs) with different concentrations to modify epoxy resins based on DGEBA. The T_g values observed decreased with the ENR content from 118 °C (neat resin) to 109 °C (20 wt % ENR). On the other hand, the impact strength increased notably with higher amounts of ENR in the blend (both notched (6.87 ± 0.8 J/m to 16.59 ± 0.6 J/m) and unnotched (1.85 ± 0.1 J/m to 2.55 ± 0.06 J/m)).

Some authors such as Imbernon et al. [94], also investigated the reprocessability of the ENRs. In this study, a disulfide function was introduced to ENR using dithiodibutyric acid (DTDB) as crosslinker. Then they compared the results with the properties of dodecanedioic acid (DA) cured samples. The mechanical properties of these cured systems are shown in Table 17.

Table 17. Mechanical properties of epoxidized natural rubber (ENR) cured with dithiodibutyric acid (DTDB) and dodecanedioic acid (DA) adapted from [94].

Sample	Stress at Break (MPa)	Young's Modulus (MPa)
ENR/DTDB	12 ± 2	1.67 ± 0.2
ENR/DA	10 ± 1	-

These authors previously published other studies demonstrating the efficiency of dicarboxilic acids as crosslinkers using 1,2-dimethylimidazole as an accelerator [95–97]. In these studies, they determined that the T_g values increased non-linearly with the amount of DA. This fact and the optimal properties of these systems, make them greener alternatives for the substitution of vulcanized natural rubber in a range of applications.

7. Epoxy Lignin Derivatives

Lignin is one of the main components of wood and has great potential as green raw material. It is considered one of the most abundant biopolymers in plants with amorphous structure and aromatic nature [98]. The combination of various functional groups in each structural unit of lignin allows for it to be used in a large number of functionalization reactions to generate high value products. The lignin can be extracted using different methods which affects the final structure of the lignin obtained [98]. The most current types of lignin depending on the extraction technology used are: Kraft

Lignin, lignosulfonates, soda lignin, organosolv lignin, klason lignin, steam explosion lignin and dilute acid lignin [99–101].

Current research in biopolymers based on lignin derivatives has aroused the interest of the scientific community in recent years. This section is focused in the most recent lignin epoxidation studies.

With the aim of substituting petroleum-based epoxy resins, Ferdosian et al. [102], synthesized lignin-based epoxy resins in a preliminary study in 2012. For the lignin epoxidation, the authors used de-polymerized organosolv lignin in alkali medium with epichlorohydrin. With this synthesis, they demonstrated the efficiency of a pre-depolymerization process before the epoxidation of lignin. This method allowed for the generation of a bio-based epoxy resin with a suitable molecular weight distribution. In other publications [101], the same authors synthesized a bio-based epoxy resin using a de-polymerized hydrolysis lignin (DHL). The de-polymerization process was performed at low pressure and temperature conditions and then the de-polymerized lignin was reacted with epichlorohydrin. Then, an exhaustive characterization of the cured samples using 4,4-diaminodiphenyl methane (DDM) as a crosslinker was performed and compared with DGEBA resins. For the mechanical characterizations fibre glass reinforced samples were produced. The mechanical properties for the different samples are shown in Table 18.

Table 18. Mechanical properties of de-polymerized hydrolysis lignin (DHL)-Epoxy—DGEBA blends cured with 4,4-diaminodiphenyl methane (DDM) adapted from [101].

Sample (% by Weight)	Tensile Strength (MPa)	Young's Modulus (GPa)	Flexural Strength (MPa)	Flexural Modulus (GPa)
100%DGEBA-DDM	214 ± 4	17.5 ± 0.4	266 ± 5	13 ± 0.3
25%DHL-Epoxy-75%DGEBA-DDM [a]	187 ± 5	23.2 ± 0.7	258 ± 4	13.2 ± 0.2
50%DHL-Epoxy-50%DGEBA-DDM	187 ± 6	18.5 ± 0.6	214 ± 4	13 ± 0.3
75%DHL-Epoxy-25%DGEBA-DDM	182 ± 3	23.1 ± 0.4	149 ± 3	10.6 ± 0.2
100%DHL-Epoxy-DDM	138 ± 4	12.3 ± 0.3	47 ± 2	5 ± 0.1

[a] A stoichiometric amount of DDM and DHL-Epoxy-DGEBA was used in all the samples.

As observed in Table 18, all the mechanical properties tend to decrease when a higher amount of DHL-Epoxy was introduced in the mixtures. The authors explain this fact as a result of the poor bonding quality of the DHL-Epoxy to the glass fibers. However, the samples containing 75% DHL-Epoxy showed higher Young's modulus and flexural modulus compared with the 100% DGEBA based samples, which suggest a potential application as a polymer matrix.

In 2014, Asada et al. [103] studied the glycidylation of low molecular weight lignin extracted from steam-exploded lignocellulosic biomass and cured the final sample using lignin as bio-based curing agent or TD2131 (a phenol novolac) as chemical curing agent. The obtained results were also compared with DGEBA based cured resins with the same curing agents. The yield range obtained for the synthesis of the lignin epoxy resins was from 63.5% to 68.2% which are good values compared with the yields of DGEBA based resins. The authors also demonstrated the thermal stability of lignin epoxy-DGEBA mixtures which makes them good alternatives for electronic applications.

Other lignin derivatives obtained from its depolymerization have also been used to generate a new range of green epoxy resins. This is the case of vanillin, a monoaromatic molecule that can be obtained from lignin. Fache et al. [104], used vanillin derivatives such as methoxyhydroquinone, vanillic acid and vanillyl alcohol to be glycidylated to obtain biobased epoxy monomers [105]. These monomers were then cured using isophorone diamine (IPDA) as curing agent. The T_g values obtained from the results are shown in Table 19.

Table 19. T_g values of vanillin derived epoxy resins and DGEBA cured with IPDA adapted from [104].

Sample	T_g (°C)
DGEBA/IPDA	166
Diglycidyl ether of vanillyl alcohol/IPDA	97
Diglycidyl ether of methoxyhydroquinone	132
Diglycidyl ether of vanillic acid	152

The T_g values obtained for the diglycidyl ether of vanillyl alcohol and diglycidyl ether of methoxyhydroquinone are quite similar to the DGEBA one. From these results, those authors were led to the conclusion that vanillin derived monomers could be used as potential alternatives to substitute DGEBA thermosetting resins in some applications, such as coatings or structural composites.

In 2017, Wang et al. 2017 [106] synthetized two novel bio-based epoxy monomers EP1 and EP2 from the lignin derivative. The vanillin-based epoxies showed T_g of 214 °C, tensile strength of 80.3 MPa, and tensile modulus of 2709 MPa, much higher than the cured DGEBA reference sample. The two systems showed also high-performances in terms of flame retardancy with UL-94 test.

Finally, Shibata et al. [107] developed a bio based aromatic epoxy resin (DGEDVCP) by the synthesis of the glycidylation of the crossed-aldol condensation product (DVCP) of vanillin and cyclopentanone. The DGEDVCP resins were cured with renewable quercetin (QC) and guaiacol novolac (GCN) and with petroleum-based phenol novolac (PN). The fully biobased epoxy resin systems utilizing renewable phenolic compounds showed a flexural strength ranging between 67 MPa and 105.9 MPa and corresponding flexural modulus between 2600 MPa and 3820 MPa, making them a possible alternative to the conventional petroleum-based epoxy resin systems.

8. Rosin Based Resin

Rosin is the major component of pine resin, which comprises approximate 70% rosin, 15% turpentine and 15% debris and water [108]. For years, rosin resin and its derivatives have been used for the production of soaps, paper sizing, printing inks, surface coatings, adhesives and rubber additives.

The two acids that compose rosin are isomeric abietic and pimaric types.

Recently, there has been a growing interest in rosin acids as feedstock chemicals for polymers or other chemical products. The recent technical development and progress are reported in a book from 2012 by Zhang [108], where the most recent developments in the utilization of rosin and terpinenes are provided. The 2D structure of rosin is shown in the following Figure 4.

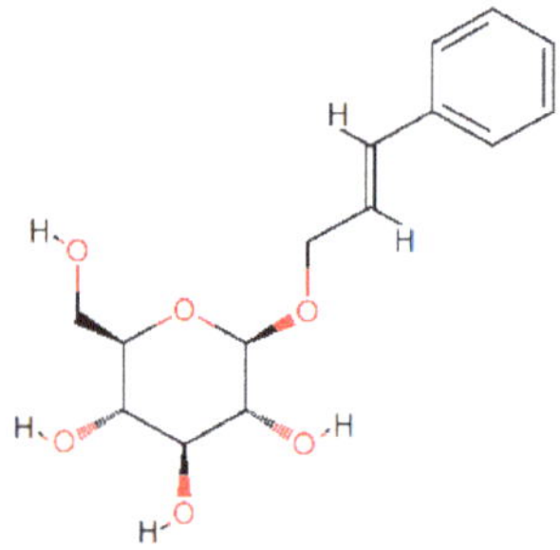

Figure 4. The 2-D structure of Rosin

Chemicals which can be derived from rosin are curing agents of anhydride type, carboxylic acid type and amine type, rosin-derived epoxies, rosin-derived monomers and surfactants.

In a work from 2012 of Liu et al. [109], a high performance bio-based epoxy was synthesized using both rosin-based epoxy monomer and rosin-based curing agent. The preparation of the resin started from a 1:1 stoichiometeric ratio of maleopimaric acid and triglycidyl ester of maleopimaric acid together with catalyst 2-ethyl-4-methylimidazole. The synthesis of maleopimaric acid was reported in detail in a previous work of the authors [110]. A glass transition temperature (T_g) of 164 °C and flexural strength and modulus, respectively, as high as 70 and 2200 MPa, were exhibited by the cured rosin-based epoxy. The impact strength and the elongation at break were, on the other end, lower than the values showed by a petroleum-based reference resin, as reported in Table 20.

Table 20. Mechanical properties of cured tirglycidyl ester of maleopimaric acid and petroleum-based DEGBA adapted from [109].

Sample	Flexural Modulus (MPa)	Flexural Strength (MPa)	Impact Strength (kJ/m^2)	Strain at Break (%)
rosin-based	2200 ± 30	70 ± 1	2.1 ± 0.2	1.9 ± 0.3
DGEBA	3000 ± 200	80 ± 3 [24]	3.2 [25]	2.6 [24]

The properties of bio-based epoxy resins derived from rosin with different flexible chains were investigated by Deng et al. [111], who obtained triglycidyl ester FPAE and glycidyl ethers FPEG1, FPEG2, and FPEG3 from rosin and studied the effects of flexible chains on tensile strength properties. Rosin based systems characterized by different ratios of FPA, Ethylene glycol diglycidyl ether (EGDE) and curing agent were synthesized. The authors also performed viscosity measurements at room temperature, showing a viscosity higher than 100 Pa s for the FPAE and FPEG1 and decreased values for FPEG2 and FPEG3, respectively of 43.5 Pa s and 7.8 Pa s.

In Table 21, the glass transition temperature T_g and the mechanical properties of products cured with different flexible chains are reported.

Table 21. T_g and mechanical properties of rosin-based resin systems cured with different flexible chains adapted from [111].

Sample	T_g (°C)	Tensile Strength (MPa)	Tensile Modulus (GPa)	Breaking Elongation (%)
E-44	140	56.25	0.29	12.35
FPAE1C	167	48.54	0.471	13.37
FPEG1C	81	68.75	0.495	17.35
FPEG2C	79	58.18	0.300	20.54
FPEG3C	75	42.41	0.270	13.67

In 2013, Li et al. [112] developed a bio-based epoxy derived from dehydroabietylamine (DHAA), which is a derivative of rosin acid and after compared with a benzylamine based epoxy. Those authors synthesized two glycidyl amine type epoxies diglycidyl dehydroabietylamine (DGDHAA) derived from DHAA, as well as a diglycidyl benzylamine (DGBA) derived from benzylamine was also obtained for further comparison. The authors investigated thermal, mechanical properties and crosslink density. The epoxy derived from dehydroabietylamine showed a higher glass transition temperature than the benzylamine-based one but lower, even if still comparable, tensile and flexural mechanical properties, explained by the authors with a weaker molecular motion ability and a lower cross-link density.

9. Summary and Discussion

A review of recent bio-based epoxy resins of different natural origins was presented, with the perspective of looking for alternative thermosets for the aviation sector, especially as matrix systems of composites for aircraft interior and secondary structures. The chemical formulation and resulting mechanical properties, glass transition temperature and viscosity have been reviewed.

Such bio-based epoxy systems were developed with the objective of obtaining renewable alternatives to petroleum based polymeric materials and performances comparable to them, although not specifically oriented to the aviation sector.

In the following section, they are reported in terms of comparison in respect to their petroleum-based counterparts or to commercially available epoxy resins, according to what is reported in each original reviewed work. Furthermore, in order to identify potentialities in the aviation sector, petrol-sourced epoxy systems currently used in this field have been added for a direct comparison of their mechanical properties. Finally, in Appendix A, summarized tables (Tables A1–A5) are presented, reporting the range of values for glass transition temperature and mechanical properties and potential applications of each resin system reviewed in this work according to the respective authors suggestions. It can be seen that many of the examined bio-thermosetting are suggested for matrix applications in composite laminates.

In particular, three comparisons are presented comprising tensile, flexural and impact mechanical properties. Figure 5 shows a comparison between the different bio-based epoxy groups, such as natural oil based (n.o.b.), isosorbide based (i.b.), furan based (f.b.), natural phenolic and polyphenolic (n.p.), epoxidized natural rubber (n.r.), rosin based (r.b.) and lignin derivatives (l.d.) and the corresponding petroleum-based counterpart (p.c.). The comparison is presented in terms of tensile modulus vs. tensile strength. Furthermore, the range of values of thermosetting resins currently used in the aerospace industry is also reported according to the recent review presented by Hamerton and Mooring [10]. Additionally, a reference value of RTM 6, a commercial aerospace grade epoxy resin [113]; widely used as matrix in composite materials is reported.

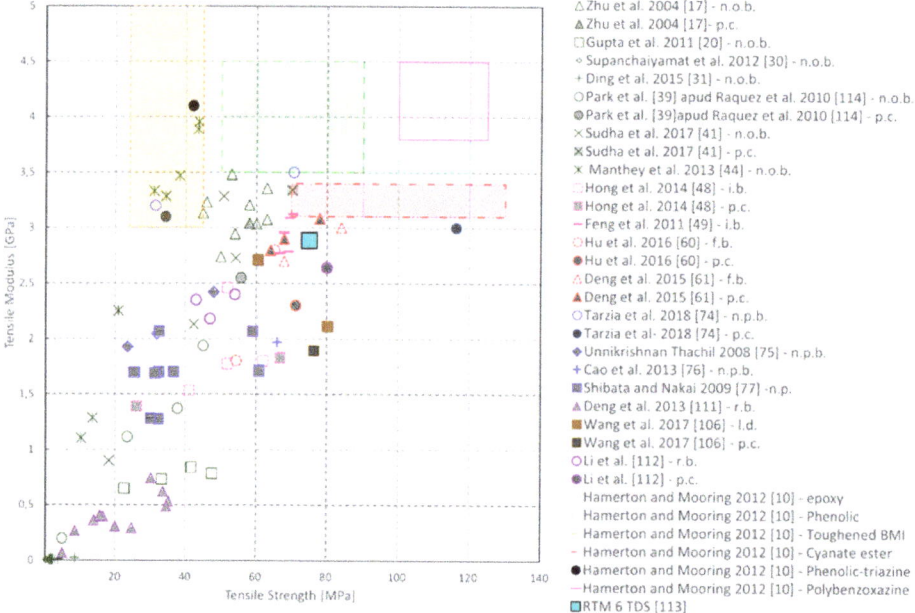

Figure 5. Tensile modulus vs. tensile strength comparison from the works of [10,17,20,30,31,39,41,44,48,49,60,61,74–77,106,111–114].

Despite the scattering of the data, due to the different bio-content, chemical formulation and curing process methods, it is possible to identify useful information about the mechanical performances of bio-resins as potential application in secondary structures composites. A cluster of points showing tensile modulus and strength comparable to the RTM 6 ones can be identified. Those points represent the renewable 2,5-furandicarboxylic acid (FDCA) presented in Deng et al. [61], the furanyl thermosetting polymers by Hu et al. [60] and isosorbide based epoxy resin by Feng et al. [49], laying in the same range of RTM 6 resin. Furthermore, interesting results characterize the epoxidized soy-based resins synthetized by Zhu et al. [17], the epoxidized hemp oil-based bioresins by Manthey et al. [44] and the epoxy resin derived from gallic acid in Tarzia et al. [74], where higher tensile modulus and relatively lower tensile strength than RTM 6 are shown. Such mechanical characteristics suggest them as suitable candidates for matrices of composite materials with enhanced modulus. Additionally, the tensile characteristics of the hemp oil-based bioresins by Manthey et al. [44] and the epoxy resin from gallic acid by Tarzia et al. [74], are located in the range values of the phenolic resin, that are generally employed in aircraft interiors, such as interior panels of glass honeycomb composites of civil airliners, flooring and partitions [10].

In the modern aerospace industry, different types of petroleum-based thermosets are employed, depending on the different performance demands, budget, aerospace sector (civil or military, smaller executive and light aircrafts) [10]. This is the case of toughened maleimide (BMI), with applications in high performance structural elements, cyanate ester, employed in space and satellite applications, polybenzoxazine, for coating and electrical applications, and epoxy for high temperature adhesives, [10] characterized by very high tensile strength and modulus. They are reported in Figure 5, according to the ranges of values presented in [10], although these applications are out of the scope of the present review on bio-based resins systems for applications in secondary and interior structures.

A comparison in terms of flexural modulus vs. flexural strength of the bio-based systems is reported in Figure 6, where bio-based epoxy synthesized from renewable gallic acid from the work of Cao et al. [76] and the soy-based epoxy resin system by Zhu et al. [17] show flexural mechanical characteristics comparable to the RTM 6 reference value [113]. Other resin systems, such as the rosin-based epoxy synthetized by Li et al. [112], the epoxy blends from castor oil by Sudha et al. [41] and the isosorbide-methacrylate based epoxy by Sadler et al. [51] present comparable or higher values in terms of flexural modulus and a lower flexural strength than the aerospace high performance resin; thus resulting more brittle than it. Finally, very interesting flexural properties characterize the isosorbide-based resin by Łukaszczyk et al. [50], with enhanced values of flexural modulus, ranging between 5.5 GPa (TETA) and 17.4 GPa (PHA), reported in detail in Appendix A, and values of flexural strength from 100.5 (THPHA) to 228.3 (TETA), making them a promising alternative to petrol-derived bisphenol A resin, as suggested by the same authors.

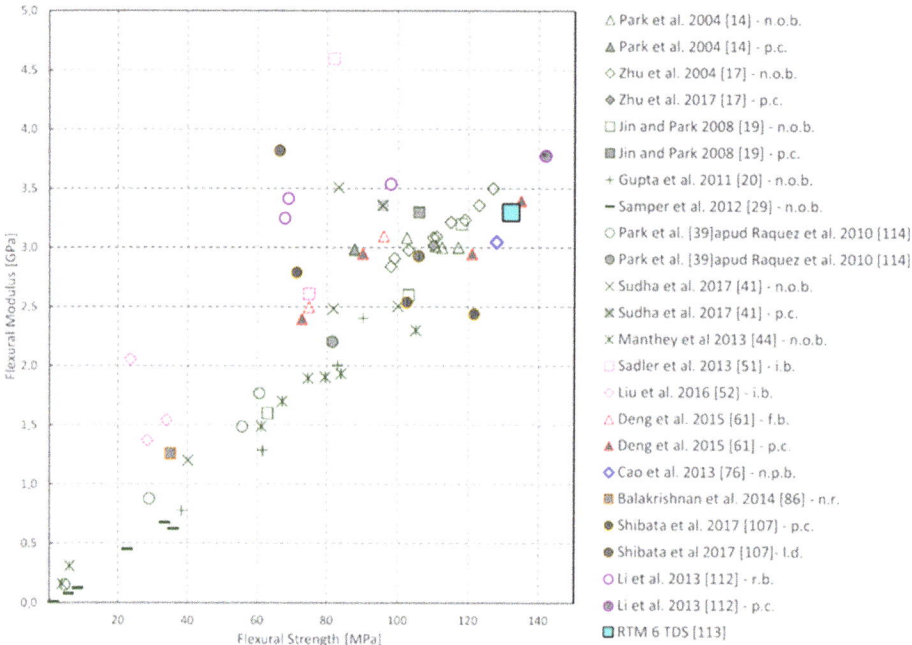

Figure 6. Flexural modulus vs. flexural strength comparison from the works of [14,17,19,20,29,39,41,44,51,52,61,76,86,107,112–114].

Further considerations can be drawn for the group of natural oil-based epoxy by examining their performances under impact loading conditions, (Figures 7 and 8). For the reported bibliography, it can be seen that the addition of a certain percentage of epoxidized natural oil content to petroleum-based systems, enhance the Charpy impact and the Izod impact strengths. In Figures 7 and 8, the gray filled symbols represent petroleum-based control samples and the unfilled symbols represent the blend of resins with a certain percentage of eco content. Such an increase of toughness with the increase of bio content in the epoxy blend is related to the change of crosslinking density of the polymer networks, which results in an increase of absorption of the energy of impact [41].

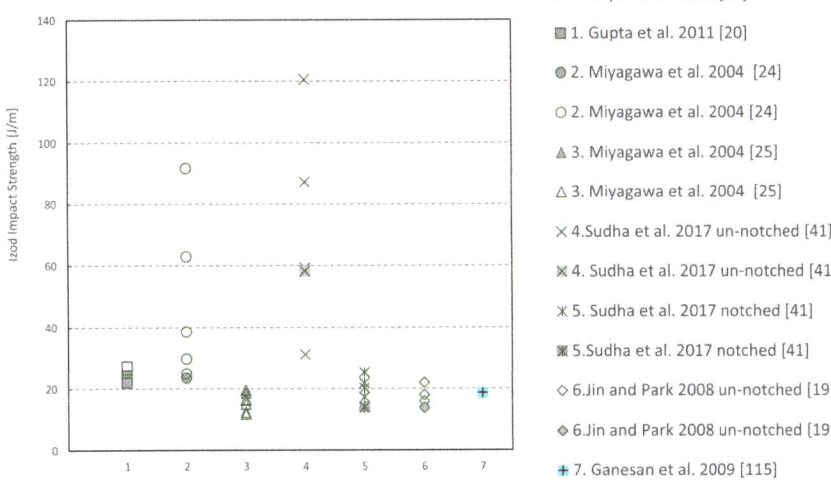

Figure 7. Izod impact strength of epoxidizes natural oil resins from the works of [19,20,24,25,41,115].

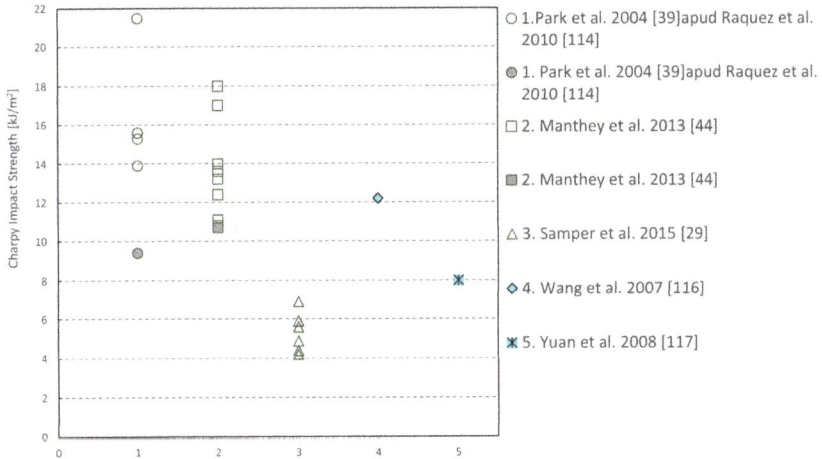

Figure 8. Charpy impact strength of epoxidized natural oil resins from the works of [29,39,44,114,116,117].

Furthermore, the bio-based resins show comparable or higher values than high-performances resins used in the aviation sector, like neat resin cyanate by Ganesan et al. [115], bismaleimidodiphenylmethane (BMI) —diallylbisphenol A (BA) by Wang et al. [116] and cyanate resin system by Wang et al. [117].

Together with the mechanical properties, for applications in composites of all the studied bio-based epoxy systems, a key aspect lays on the adhesion with the fibers for composite materials. In fact, besides the mechanical properties of the single components (epoxy resin matrix and fibers), when they are employed to produce composite materials, a good adhesion contributes to an improvement of the mechanical characteristics of the composite.

Such an aspect is also related to other factors, such as the viscosity of the resin systems to properly permeate the fibers. In Table A6 of Appendix B, the values determined by rheological analyses characterizing the bio-based systems are summarized in terms of kinematic and dynamic viscosity. In order to provide a first comparison with the resins generally used in the aerospace for the production of composite, the values characterizing the RTM 6 resin are also reported [113]. Here, it comes clear that bio-based systems such as the isosorbide-based systems presented in [51,52] and natural phenolic based resin systems of glycerol polyglycidyl ether (GPE) and sorbitol polyglycidyl ether (SPE) cured with tannic acid (TA) presented in [77], show interesting properties in terms of viscosity, making them promising for fibers permeation.

The studies reported in this work already provide several examples where the mechanical properties and viscosity of the systems are suitable for applications in fiber composites and make them promising candidates for interior components of an airplane, even if more research is still required. Despite significant advances on the use of bio-based epoxy systems, there is still a need to further optimize their performances to make them competitive for applications in the aviation sector.

Even though it is outside the scope of the present review, it is worth mentioning that initial studies on the costs of bio-based resins and natural fibers for composites and life cycle assessment analysis has been carried in [118,119]. Further studies must still be conducted as most of the abovementioned resin systems are not yet at a commercial production level.

Funding: This research was funded by the European Union's Horizon 2020 research and innovation program under grant number 690638.

Conflicts of Interest: The authors declare no conflict of interest.

Appendix A

Table A1. Natural oil-based epoxy systems: chemical composition, glass transition temperature, mechanical properties, potential applications.

Nature	Epoxy System	Sample Description	T_g (°C)	Tensile Strength (MPa)	Tensile Modulus (MPa)	Flexural Strength (MPa)	Flexural Modulus (MPa)	Potential Applications Suggested by the Authors	References
Natural oil-based epoxy	Epoxidized soybean oil (ESO)	Commercial ESO (from 30 to 10 wt %) mixed with Shell Epon 9500 epoxy resin	61.9–72.3 respectively	60–51 respectively	3193–2807 respectively	99–119 respectively	2910–3234 respectively	Composites (enhancement of mechanical properties needed)	Zhu et al. [17]
		Epoxidized allyl soyate (from 30 to 10 wt %) mixed with Shell Epon 9500	65.0–75.1 respectively	54–53 respectively	2952–2972 respectively	103–127 respectively	2979–3503 respectively	-	Zhu et al. [17]
		ESO (from 0 to 100 wt %)-DGEBA blends	108–57 respectively	-	-	-	-	-	Altuna et al. [18]
		ESO (from 0 to 60 wt %)-DGEBA blends	-	-	-	106–63 respectively	3300–1600 respectively	-	Jin and Park [19]
	Epoxidized linseed oil (ELO)	ELO-bio-based long chain diacid (Pripol 1009)	-	1.65	≈7.3–7.9	-	-	Composites, laminates, adhesives	Supanchaiyamat et al. [30]
		ELO-Adipic acid	1.5 (DMA)	8.8	22	-	-	-	Ding et al. [31]
		ELO-methyl nadic anhydride reinforced with slate fibres with differed silane treatments	-	328.2–359.1	21,900–25,600	299.2–402.1	18,400–19,700	-	Samper et al. [29]
	Epoxidized canola oil (ECO)	ECO-phthalic anhydride (PA)	−24.1–16.1 depending on the curing temperature and PA proportions	-	-	-	-	For making lignocellulosic fibre- and particle-based biocomposites	Omonov and Curtis [37]
	Epoxidized castor oil	Epoxidized castor oil-DGEBA Curing catalyst: N-benzylpyrazinium hexafluoroantimonate (BPH) (99:1 wt ratio)	197–38 (0 to 100 wt % epoxidized castor oil)	-	-	82.5	3400	-	Park et al. [38]
		Epoxidized castor oil-DGEBA Curing catalyst: N-benzylquinoxalinium hexafluoroantimonate (BQH) (99:1 wt ratio)	-	-	-	122.8	2800	-	Park et al. [39]
		Epoxidized castor oil (0–50 wt %) DGEBA Curing agent: TETA	96.64–39.21 respectively	70.18–18.26 respectively	343.11–900.59 respectively	95.644–40.04 respectively	3358.05–1200.79 respectively	-	Sudha et al. [41]
	Epoxidized karanja oil (KO)	Epoxidized KO Curing agents: CA and TA	112.70 (CA) 108.64 (TA)	10.60 (CA) 4.50 (TA)	2.65 (CA) 2.58 (TA)	-	-	Coatings and lamination	Kadam et al. [42]

Table A2. Isosorbide- and furan-based epoxy systems: chemical composition, glass transition temperature, mechanical properties, potential applications.

Nature	Epoxy System	Sample Description	T_g (°C)	Tensile Strength (MPa)	Tensile Modulus (MPa)	Flexural Strength (MPa)	Flexural Modulus (MPa)	Potential Applications Suggested by the Authors	References
Isosorbide based epoxy	Diglicidyl eter of isosorbide (DGEI)	DGEI Curing agents: DETA and ISODA	76 (DETA) 43 (ISODA)	62 (DETA) 41 (ISODA)	1798 (DETA) 1532 (ISODA)	-	4027 (DETA) 1168 (ISODA)	Replacement of BPA (for food contact applications). Industry additives, can coatings, biomedical applications like bone cements and drug delivery systems, packaging, automotive industry.	Hong et al.
		DGEI Curing agents: PHA, THPHA, TETA and IPHA	108 (PHA) 95 (THPHA) 49 (TETA) 73 (IPHA)	-	-	225.5 (PHA) 100.5 (THPHA) 228.3 (TETA) 158.8 (IPHA)	17,400 (PHA) 15,100 (THPHA) 5500 (TETA) 14,600 (IPHA)		Łukaszczyk et al. [37]
	Bisisosorbide diglicidyl eter	Bisisosorbide diglydicyl eter Curing agent: Jeffamine T403	48 but can be increased to 200 °C changing the curing agent	68.8	2944	-	-		Feng et al. [49]
Furan based epoxy	BOF and BOB	BOF/BOB—DGEBA Curing agents: PACM and EPIKURE W	80 to 150 depending on the proportions and curing agent.	-	-	-	-	Adhesives, structural and engineering materials and composites	Hu et al. [58]
		BOF Curing agents: DFDA, CH3-DCBA, PACM	69 (DFDA) 62 (CH3-DCBA) 72 (PACM)	-	-	-	-		Hu et al. [59]
	DGF	DGF Curing agents: MHHPA, D230	152 (MHHPA) 101.2 (D230)	84 (MHHPA) 68 (D230)	3000 (MHHPA) 2700 (D230)	96 (MHHPA) 75 (D230)	3100 (MHHPA) 2500 (D230)		Deng et al. [61]

Table A3. Phenolic and polyphenolic epoxy: chemical composition, glass transition temperature, mechanical properties, potential applications.

Nature	Epoxy System	Sample Description	T_g (°C)	Tensile Strength (MPa)	Tensile Modulus (MPa)	Flexural Strength (MPa)	Flexural Modulus (MPa)	Potential Applications Suggested by the Authors	References
Phenolic and polyphenolic epoxy	GEHDGTE, GEFDGTE, GEC	GEGTE, GEC Curing agent IPDA	142 (GEGTE) 179 (GEC)	-	-	-	-		Benyahya et al. [63]
		GEHDGTE, GEFDGTE, GEC Curing agent: Lignin derivative	178 (GEC) 155 (GEFDGTE) 173 (GEFHDGTE)	-	-	63 (GEC) 56 (GEFDGTE) 40 (GEFHDGTE)	-	Electronic applications, composites	Basnet et al. [66]
	GEGA	GEGA Curing agents: IPDA, DPG, BLMA	158 (IPDA) 98 (DPG) 136 (BDMA)	43.1 (IPDA) 70.6 (DPG) 31.2 (BDMA)	3600 (IPDA) 3500 (DPG) 3200 (BDMA)	-	-		Tarzia et al. [74]
	TA	GPE, SPE Curing agent: TA	87.3 (GPE) 106.6 (SPE)	36.7 (GPE) 60.6 (SPE)	2400 (GPE) 1710 (SPE)	-	-		Shibata et al. [77]
		Cardanol based resol-DGEBA Curing agent: Amine catalyst or an acid catalyst	-	12	864	-	-		Maffezzoli et al. [80]
		BPA-Cardanol epoxy (80:20 and 50:50)	-	31.7 (80:20) 23.5 (50:50)	2045 (80:20) 1926 (50:50)	80.8 (80:20) 71.45 (50:50)	-		Unnikrishnan et al. [75]
		NC-514 Curing agents: IPDA, Jeffamine D400	50 (IPDA) 15 (Jeffamine D400)	-	-	-	-	Composites, binders and coatings	Jaillet et al. [81]
	Cardanol epoxy	NC-514-Sorbitol/Isosorbide epoxies Curing agents: IPDA, Jeffamine T403	83 (25/75 Epoxidized cardanol/Epoxidized isosorbide cured with IPDA) 60 (50 wt % Epoxidized cardanol/Epoxidized sorbitol cured with IPDA) 37 (50 wt % Epoxidized cardanol/Epoxidized sorbitol cured with Jeffamine T403)	-	-	-	-		Darroman et al. [82]
		CNE Curing agent: CPA	50–84	-	-	-	-		Atta et al. [83]

Table A4. Epoxidized Natural rubber (ENR) and epoxy lignin derivatives: chemical composition, glass transition temperature, mechanical properties, potential applications.

Nature	Epoxy System	Sample Description	T_g (°C)	Tensile Strength (MPa)	Tensile Modulus (MPa)	Flexural Strength (MPa)	Flexural Modulus (MPa)	Potential Applications Suggested by the Authors	References
Epoxidized Natural rubber (ENR)	ENR	ENR-DGEBA Curing agent: Nadic methyl anhydride (K 68)	112 (5 wt % ENR)–109 (20 wt % ENR)	-	-	-	-	-	Mathew et al. [99]
		ENR Curing agent: DTDB	-	12	1.67	-	-	-	Imbernon et al. [98]
	Depolymerized lignin epoxy	DHL epoxy-DGEBA Curing agent: DDM	-	138 (100% DHL)–187 (25% DHL)	12,300 (100% DHL) 23,200 (25% DHL)	47 (100% DHL)–258 (25% DHL)	5000 (100% DHL)–13,200 (25% DHL)	Electronics, substitute for fossil resource-derived bisphenol A, polymer matrix for manufacture of bio-based fibre-reinforced plastics or composites	Ferdosian et al. [100]
Epoxy lignin derivatives	Vanillin derivatives	Diglycidyl ethers of vanillyl alcohol, vanillic acid and methoxyhydroquinone Curing agent: IPDA	97 (diglycidyl ether of vanillyl alcohol) 132 (Diglycidyl ether of methoxyhydroquinone) 152 (Diglycidyl ether of vanillic acid)	-	-	-	-		Fache et al. [104]

Table A5. Rosin based epoxy: chemical composition, glass transition temperature, mechanical properties, potential applications.

Nature	Epoxy System	Sample Description	T_g (°C)	Tensile Strength (MPa)	Tensile Modulus (MPa)	Flexural Strength (MPa)	Flexural Modulus (MPa)	Potential Applications Suggested by the Authors	References
Rosin based	Triglycidyl ester FPAE and glycidyl ethers FPEG1, FPEG2, and FPEC3 obtained from Rosin	E-44 FPAE1C FPEG1C FPEG2C FPEG3C	140 167 81 79 75	56.25 48.54 68.75 58.18 42.41	290 471 495 300 270	-	-	-	Deng et al. [111]
	Rosin based-epoxy monomer and curing agent	Maleopimaric acid (MPA) and triglycidyl ester of maleopimaric acid	164	-	-	70	2200		Liu et al. [109]
	Rosin-based Epoxy Monomer	Two glycidyl amine type epoxies: diglycidyl dehydroabietylamine (DGDHAA) derived from DHAA (rosin) and diglycidyl benzylamine (DGBA) derived from benzylamine hexahydrophthalic anhydride HHPA		47 43 54	2180 2330 2400	68 69 98	3250 3417 3540		Li et al. [112]

Appendix B

Table A6. Viscosity of the reviewed bio-based epoxy resin systems and reference value for RTM 6 resin [113].

Type of Measurement	Type of Bio-Based Resin System	Range of Values	Temperature of Measurements	Reference
Kinematic viscosity	ESO	170.87 mm^2·s^{-1}—20.41 mm^2·s^{-1}	40 °C at 100°C	Erhan et al. [5]
Dynamic viscosity	ELO with different amine catalysts	400 mPa s—2000 mPa s depending on the catalyst	at 140°C	Supanchaiyamat et al. [8]
Dynamic viscosity	castor oil/DGEBA blends at various wt %—TETA as curing agent	950 mPa s—1050 mPa s (initial)	20 °C	Sudha et al. [41]
Dynamic viscosity	EHO	845 mPa s	25 °C	Manthey et al. [44]
Dynamic viscosity	Pollit/MMA (70/30) Pollit/TPGDA (70/30) Tribest MMSO	500 mPa s 13,900 mPa s 5700 mPa s 1200 mPa s	24 °C	Åkesson et al. [45]
Dynamic viscosity	Isosorbide-based DGEI/ISODA	<10,000 mPa s (initial)	25 °C	Hong et al. [48]
Kinematic viscosity	IS-EPO	60,120 mPa s	20 °C	Lukaszczyk et al. [50]
Dynamic viscosity	neat IM	156 mPa s	25 °C	Sadler et al. [51]
Dynamic viscosity	AESO IM-AESO IM-MAESO IM	4789 ± 69 mPa s 151 ± 1 mPa s 186 ± 7 mPa s 12 ± 1 mPa s	30 °C	Liu et al. [52]
Dynamic viscosity	GEGA	2000 mPa s	room temperature	Tarzia et al. [74]
Dynamic viscosity	BPA/cardanol epoxy 80:20 BPA/cardanol epoxy 50:50	10,485 mPa s 9868 mPa s	25 °C	Unnikrishnan et al. [75]
Dynamic viscosity	GPE SPE	150 mPa s 5000 mPa s	25 °C	Shibata et al. [77]
Dynamic viscosity	Resole–epoxy Resole–epoxy Resole–epoxy Resole–epoxy-acid catalyst	470 mPa s 2800 mPa s 4200 mPa s 4000 mPa s	25 °C	Maffezzoli et al. [80]
Dynamic viscosity	cardanol novolac epoxy (CNE) resin cardanol polyamine hardener (CPA)	1150 mPa s 2800 mPa s	-	Atta et al. [83]
Dynamic viscosity	triglycidyl ester FPAE glycidyl ethers from rosin FPEG1 glycidyl ethers from rosin FPEG2 glycidyl ethers from rosin FPEG3	>100,000 mPa s >100,000 mPa s 43,500 mPa s 7800 mPa s	25 °C	Deng et al. [111]
Dynamic viscosity	RTM 6	32–38 mPa s (initial) 59–89 mPa s (after 2 h)	120 °C	RTM 6-TDS [113]

References

1. Pascault, J.; Sautereau, H.; Verdu, J.; Williams, R.J.J. *Thermosetting Polymers*; Marcel Dekker: New York, NY, USA, 2002; ISBN 0824706706.
2. Holbery, J.; Houston, D. Natural-fibre-reinforced polymer composites in automotive applications. *J. Miner. Met. Mater. Soc.* **2006**, *58*, 80–86. [CrossRef]
3. Prolongo, S.G.; Gude, M.R.; Sanchez, J.; Ureña, A. Nanoreinforced epoxy adhesives for aerospace industry. *J. Adhes.* **2009**, *85*, 180–199. [CrossRef]
4. Guo, C.; Zhou, L.; Lv, J. Effects of expandable graphite and modified ammonium polyphosphate on the flame-retardant and mechanical properties of wood flour-polypropylene composites. *Polym. Polym. Compos.* **2013**, *21*, 449–456. [CrossRef]
5. Sharmin, E.; Alam, M.S.; Philip, R.K.; Ahmad, S. Linseed amide diol/DGEBA epoxy blends for coating applications: Preparation, characterization, ageing studies and coating properties. *Prog. Org. Coat.* **2010**, *67*, 170–179. [CrossRef]
6. Gibson, R.F. A review of recent research on mechanics of multifunctional composite materials and structures. *Compos. Struct.* **2010**, *92*, 2793–2810. [CrossRef]
7. Jin, F.L.; Li, X.; Park, S.J. Synthesis and application of epoxy resins: A review. *J. Ind. Eng. Chem.* **2015**, *29*, 1–11. [CrossRef]
8. Shen, L.; Haufe, J.; Patel, M. *Product Overview and Market Projection of Emerging Biobased Plastics (PROBIP 2009)*; Utrecht University: Utrecht, The Netherlands, 2009.
9. Soutis, C. Fibre reinforced composites in aircraft construction. *Prog. Aerosp. Sci.* **2005**, *41*, 143–151. [CrossRef]
10. Hamerton, I.; Mooring, L. The use of thermosets in aerospace applications. *Thermosets Struct. Prop. Appl.* **2012**, 189–227. [CrossRef]
11. Yang, Y.; Boom, R.; Irion, B.; van Heerden, D.J.; Kuiper, P.; de Wit, H. Recycling of composite materials. *Chem. Eng. Process. Process Intensif.* **2012**, *51*, 53–68. [CrossRef]
12. Soutis, C. Carbon fiber reinforced plastics in aircraft construction. *Mater. Sci. Eng. A* **2005**, *412*, 171–176. [CrossRef]
13. Kausar, A. Role of Thermosetting Polymer in Structural Composite. *Am. J. Polym. Sci. Eng.* **2017**, *5*, 1–12.
14. Park, S.J.; Jin, F.L.; Lee, J.R. Thermal and mechanical properties of tetrafunctional epoxy resin toughened with epoxidized soybean oil. *Mater. Sci. Eng. A* **2004**, *374*, 109–114. [CrossRef]
15. Adhvaryu, A.; Erhan, S.Z. Epoxidized soybean oil as a potential source of high-temperature lubricants. *Ind. Crops Prod.* **2002**, *15*, 247–254. [CrossRef]
16. Petrović, Z.S.; Zlatanić, A.; Lava, C.C.; Sinadinović-Fišer, S. Epoxidation of soybean oil in toluene with peroxoacetic and peroxoformic acids—Kinetics and side reactions. *Eur. J. Lipid Sci. Technol.* **2002**, *104*, 293–299. [CrossRef]
17. Zhu, J.; Chandrashekhara, K.; Flanigan, V.; Kapila, S. Curing and mechanical characterization of a soy-based epoxy resin system. *J. Appl. Polym. Sci.* **2004**, *91*, 3513–3518. [CrossRef]
18. Altuna, F.I.; Espósito, L.H.; Ruseckaite, R.A.; Stefani, P.M. Thermal and mechanical properties of anhydride-cured epoxy resins with different contents of biobased epoxidized soybean oil. *J. Appl. Polym. Sci.* **2011**, *120*, 789–798. [CrossRef]
19. Jin, F.-L.; Park, S.-J. Impact-strength improvement of epoxy resins reinforced with a biodegradable polymer. *Mater. Sci. Eng. A* **2008**, *478*, 402–405. [CrossRef]
20. Gupta, A.P.; Ahmad, S.; Dev, A. Modification of novel bio-based resin-epoxidized soybean oil by conventional epoxy resin. *Polym. Eng. Sci.* **2011**, *51*, 1087–1091. [CrossRef]
21. Tan, S.G.; Chow, W.S. Curing characteristics and thermal properties of Epoxidized soybean oil based thermosetting resin. *Am. Oil Chem. Soc.* **2011**, *88*, 915–923. [CrossRef]
22. Cavusoglu, J.; Çayli, G. Polymerization reactions of epoxidized soybean oil and maleate esters of oil-soluble resoles. *J. Appl. Polym. Sci.* **2015**, *132*, 1–6. [CrossRef]
23. Tsujimoto, T.; Takayama, T.; Uyama, H. Biodegradable Shape Memory Polymeric Material from Epoxidized Soybean Oil and Polycaprolactone. *Polymers* **2015**, *7*, 2165–2174. [CrossRef]
24. Miyagawa, H.; Mohanty, A.K.; Misra, M.; Drzal, L.T. Thermo-physical and impact properties of epoxy containing epoxidized linseed oil, 1: Anhydride-cured epoxy. *Macromol. Mater. Eng.* **2004**, *289*, 629–635. [CrossRef]

25. Miyagawa, H.; Mohanty, A.K.; Misra, M.; Drzal, L.T. Thermo-Physical and Impact Properties of Epoxy Containing Epoxidized Linseed Oil, 2. *Macromol. Mater. Eng.* **2004**, *289*, 636–641. [CrossRef]
26. Kanno, S.; Kawamura, Y.; Mutsuga, M.; Tanamoto, K. Determination of Epoxidized Soybean Oil and Linseed Oil in Wrapping Film and Cap Sealing. *J. Food Hyg. Soc. Jpn. (Shokuhin Eiseigaku Zasshi)* **2006**, *47*, 89–94. [CrossRef]
27. Sánchez, N.; Chirinos, J. Estabilizantes térmicos alternativos para el PVC. *Rev. Iberoam. Polímeros* **2014**, *15*, 178–197.
28. Espín, J.C.; Soler-Rivas, C.; Wichers, H.J. Characterization of the total free radical scavenger capacity of vegetable oils and oil fractions using 2,2-diphenyl-1-picrylhydrazyl radical. *J. Agric. Food Chem.* **2000**, *48*, 648–656. [CrossRef] [PubMed]
29. Samper, M.D.; Petrucci, R.; Sánchez-Nacher, L.; Balart, R.; Kenny, J.M. New environmentally friendly composite laminates with epoxidized linseed oil (ELO) and slate fiber fabrics. *Compos. Part B Eng.* **2015**, *71*, 203–209. [CrossRef]
30. Supanchaiyamat, N.; Shuttleworth, P.S.; Hunt, A.J.; Clark, J.H.; Matharu, A.S. Thermosetting resin based on epoxidised linseed oil and bio-derived crosslinker. *Green Chem.* **2012**, *14*, 1759–1765. [CrossRef]
31. Ding, C.; Shuttleworth, P.S.; Makin, S.; Clark, J.H.; Matharu, A.S. New insights into the curing of epoxidized linseed oil with dicarboxylic acids. *Green Chem.* **2015**, *17*, 4000–4008. [CrossRef]
32. Pin, J.M.; Sbirrazzuoli, N.; Mija, A. From epoxidized linseed oil to bioresin: An overall approach of epoxy/anhydride cross-linking. *ChemSusChem* **2015**, *8*, 1232–1243. [CrossRef] [PubMed]
33. Samper, M.D.; Fombuena, V.; Boronat, T.; García-Sanoguera, D.; Balart, R. Thermal and Mechanical Characterization of Epoxy Resins (ELO and ESO) Cured with Anhydrides. *J. Am. Oil Chem. Soc.* **2012**, *89*, 1521–1528. [CrossRef]
34. Pérez, J.D.E.; Haagenson, D.M.; Pryor, S.W.; Ulven, C.A.; Wiesenborn, D.P. Production and Characterization of Epoxidized Canola Oil. *Trans. ASABE* **2009**, *52*, 1289–1297. [CrossRef]
35. Campanella, A.; Fahimian, M.; Wool, R.P.; Raghavan, J. Synthesis and Rheology of Chemically Modified Canola Oil. *J. Biobased Mater. Bioenergy* **2009**, *3*, 91–99. [CrossRef]
36. Mungroo, R.; Pradhan, N.C.; Goud, V.V.; Dalai, A.K. Epoxidation of Canola Oil with Hydrogen Peroxide Catalyzed by Acidic Ion Exchange Resin. *J. Am. Oil Chem. Soc.* **2008**, *85*, 887–896. [CrossRef]
37. Omonov, T.S.; Curtis, J.M. Biobased epoxy resin from canola oil. *J. Appl. Polym. Sci.* **2014**, *131*. [CrossRef]
38. Park, S.-J.; Seo, M.-K.; Lee, J.-R.; Lee, D.-R. Studies on epoxy resins cured by cationic latent thermal catalysts: The effect of the catalysts on the thermal, rheological, and mechanical properties. *J. Polym. Sci. Part A Polym. Chem.* **2001**, *39*, 187–195. [CrossRef]
39. Park, S.-J.; Jin, F.-L.; Lee, J.-R. Effect of Biodegradable Epoxidized Castor Oil on Physicochemical and Mechanical Properties of Epoxy Resins. *Macromol. Chem. Phys.* **2004**, *205*, 2048–2054. [CrossRef]
40. Park, S.-J.; Jin, F.-L.; Lee, J.-R. Synthesis and Thermal Properties of Epoxidized Vegetable Oil. *Macromol. Rapid Commun.* **2004**, *25*, 724–727. [CrossRef]
41. Sudha, G.S.; Kalita, H.; Mohanty, S.; Nayak, S.K. Biobased epoxy blends from epoxidized castor oil: Effect on mechanical, thermal, and morphological properties. *Macromol. Res.* **2017**, *25*, 420–430. [CrossRef]
42. Kadam, A.; Pawar, M.; Yemul, O.; Thamke, V.; Kodam, K. Biodegradable biobased epoxy resin from karanja oil. *Polymer* **2015**, *72*, 82–92. [CrossRef]
43. Stemmelen, M.; Pessel, F.; Lapinte, V.; Caillol, S.; Habas, J.-P.; Robin, J.-J. A fully biobased epoxy resin from vegetable oils: From the synthesis of the precursors by thiol-ene reaction to the study of the final material. *J. Polym. Sci. Part A Polym. Chem.* **2011**, *49*, 2434–2444. [CrossRef]
44. Manthey, N.W.; Cardona, F.; Francucci, G.; Aravinthan, T. Thermo-mechanical properties of epoxidized hemp oil-based bioresins and biocomposites. *J. Reinf. Plast. Compos.* **2013**, *32*, 1444–1456. [CrossRef]
45. Åkesson, D.; Skrifvars, M.; Lv, S.; Shi, W.; Adekunle, K.; Seppälä, J.; Turunen, M. Preparation of nanocomposites from biobased thermoset resins by UV-curing. *Prog. Org. Coat.* **2010**, *67*, 281–286. [CrossRef]
46. Flèche, G.; Huchette, M. Isosorbide. Preparation, Properties and Chemistry. *Starch* **1986**, *38*, 26–30. [CrossRef]
47. Rose, M.; Palkovits, R. Isosorbide as a Renewable Platform chemical for Versatile Applications—Quo Vadis? *ChemSusChem* **2012**, *5*, 167–176. [CrossRef] [PubMed]
48. Hong, J.; Radojčić, D.; Ionescu, M.; Petrović, Z.S.; Eastwood, E. Advanced materials from corn: Isosorbide-based epoxy resins. *Polym. Chem.* **2014**, *5*, 5360–5368. [CrossRef]

49. Feng, X.; East, A.J.; Hammond, W.B.; Zhang, Y.; Jaffe, M. Overview of advances in sugar-based polymers. *Polym. Adv. Technol.* **2011**, *22*, 139–150. [CrossRef]

50. Łukaszczyk, J.; Janicki, B.; Kaczmarek, M. Synthesis and properties of isosorbide based epoxy resin. *Eur. Polym. J.* **2011**, *47*, 1601–1606. [CrossRef]

51. Sadler, J.M.; Nguyen, A.-P.T.; Toulan, F.R.; Szabo, J.P.; Palmese, G.R.; Scheck, C.; Lutgen, S.; La Scala, J.J. Isosorbide-methacrylate as a bio-based low viscosity resin for high performance thermosetting applications. *J. Mater. Chem. A* **2013**, *1*, 12579. [CrossRef]

52. Liu, W.; Xie, T.; Qiu, R. Biobased Thermosets Prepared from Rigid Isosorbide and Flexible Soybean Oil Derivatives. *ACS Sustain. Chem. Eng.* **2017**. [CrossRef]

53. Gandini, A. Furans as offspring of sugars and polysaccharides and progenitors of a family of remarkable polymers: a review of recent progress. *Polym. Chem.* **2010**, *1*, 245–251. [CrossRef]

54. Spillman, P.J.; Pollnitz, A.P.; Liacopoulos, D.; Pardon, K.H.; Sefton, M.A. Formation and Degradation of Furfuryl Alcohol, 5-Methylfurfuryl Alcohol, Vanillyl Alcohol, and Their Ethyl Ethers in Barrel-Aged Wines. *J. Agric. Food Chem.* **1998**, *46*, 657–663. [CrossRef] [PubMed]

55. Lamminpää, K.; Ahola, J.; Tanskanen, J. Kinetics of Xylose Dehydration into Furfural in Formic Acid. *Ind. Eng. Chem. Res.* **2012**, *51*, 6297–6303. [CrossRef]

56. Lamminpää, K. Formic Acid Catalysed Xylose Dehydration into Furfural. Ph.D. Thesis, University of Oulu, Oulu, Finland, 2015.

57. Cho, J.K.; Lee, J.-S.; Jeong, J.; Kim, B.; Kim, B.; Kim, S.; Shin, S.; Kim, H.-J.; Lee, S.-H. Synthesis of carbohydrate biomass-based furanic compounds bearing epoxide end group(s) and evaluation of their feasibility as adhesives. *J. Adhes. Sci. Technol.* **2013**, *27*, 2127–2138. [CrossRef]

58. Hu, F.; La Scala, J.J.; Sadler, J.M.; Palmese, G.R. Synthesis and Characterization of Thermosetting Furan-Based Epoxy Systems. *Macromolecules* **2014**, *47*, 3332–3342. [CrossRef]

59. Hu, F.; Yadav, S.K.; La Scala, J.J.; Sadler, J.M.; Palmese, G.R. Preparation and Characterization of Fully Furan-Based Renewable Thermosetting Epoxy-Amine Systems. *Macromol. Chem. Phys.* **2015**, *216*, 1441–1446. [CrossRef]

60. Hu, F.; Yadav, S.K.; Sharifi, M.; La Scala, J.; Sadler, J.; McAninch, I.; Palmesea, G. Characterization of Furanyl Thermosetting Polymers with Superior Mechanical Properties and High-Temperature Char Yield. In Proceedings of the International SAMPE Technical Conference, Long Beach, CA, USA, 23–26 May 2016.

61. Deng, J.; Liu, X.; Li, C.; Jiang, Y.; Zhu, J. Synthesis and properties of a bio-based epoxy resin from 2,5-furandicarboxylic acid (FDCA). *RSC Adv.* **2015**, *5*, 15930–15939. [CrossRef]

62. Abbas, M.; Saeed, F.; Anjum, F.M.; Afzaal, M.; Tufail, T.; Bashir, M.S.; Ishtiaq, A.; Hussain, S.; Suleria, H.A.R. Natural polyphenols: An overview. *Int. J. Food Prop.* **2017**, *20*, 1689–1699. [CrossRef]

63. Pizzi, A. Monomers, Polymers and Composites from Renewable Resources. *Monomers Polym. Compos. Renew. Resour.* **2008**, 179–199. [CrossRef]

64. Nouailhas, H.; Aouf, C.; Le Guerneve, C.; Caillol, S.; Boutevin, B.; Fulcrand, H. Synthesis and properties of biobased epoxy resins. Part 1. Glycidylation of flavonoids by epichlorohydrin. *J. Polym. Sci. Part A Polym. Chem.* **2011**, *49*, 2261–2270. [CrossRef]

65. Benyahya, S.; Aouf, C.; Caillol, S.; Boutevin, B.; Pascault, J.P.; Fulcrand, H. Functionalized green tea tannins as phenolic prepolymers for bio-based epoxy resins. *Ind. Crops Prod.* **2014**, *53*, 296–307. [CrossRef]

66. Basnet, S.; Otsuka, M.; Sasaki, C.; Asada, C.; Nakamura, Y. Functionalization of the active ingredients of Japanese green tea (Camellia sinensis) for the synthesis of bio-based epoxy resin. *Ind. Crops Prod.* **2015**, *73*, 63–72. [CrossRef]

67. Haslam, E.; Cai, Y. Plant polyphenols (Vegetable tannins): Gallic acid metabolism. *Nat. Prod. Rep.* **1994**, *11*, 41–66. [CrossRef] [PubMed]

68. Badhani, B.; Sharma, N.; Kakkar, R. Gallic acid: A versatile antioxidant with promising therapeutic and industrial applications. *RSC Adv.* **2015**, *5*, 27540–27557. [CrossRef]

69. Al, M.L.; Daniel, D.; Moise, A.; Bobis, O.; Laslo, L.; Bogdanov, S. Physico-chemical and bioactive properties of different floral origin honeys from Romania. *Food Chem.* **2009**, *112*, 863–867. [CrossRef]

70. Samanidou, V.; Tsagiannidis, A.; Sarakatsianos, I. Simultaneous determination of polyphenols and major purine alkaloids in Greek Sideritis species, herbal extracts, green tea, black tea, and coffee by high-performance liquid chromatography-diode array detection. *J. Sep. Sci.* **2012**, *35*, 608–615. [CrossRef] [PubMed]

71. Schmitzer, V.; Slatnar, A.; Veberic, R.; Stampar, F.; Solar, A. Roasting Affects Phenolic Composition and Antioxidative Activity of Hazelnuts (*Corylus avellana* L.). *J. Food Sci.* **2011**, *76*. [CrossRef] [PubMed]
72. Tomita, H.; Yonezawa, K. Epoxy Resin and Process for Preparing the Same. U.S. Patent No. 4,540,802, 10 September 1985.
73. Aouf, C.; Lecomte, J.; Villeneuve, P.; Dubreucq, E.; Fulcrand, H. Chemo-enzymatic functionalization of gallic and vanillic acids: Synthesis of bio-based epoxy resins prepolymers. *Green Chem.* **2012**, *14*, 2328–2336. [CrossRef]
74. Tarzia, A.; Montanaro, J.; Casiello, M.; Annese, C.; Nacci, A.; Maffezzoli, A. Synthesis, Curing, and Properties of an Epoxy Resin Derived from Gallic Acid. *BioResources* **2017**, *13*, 632–645. [CrossRef]
75. Unnikrishnan, K.P.; Thachil, E.T. Synthesis and characterization of cardanol-based epoxy systems. *Des. Monomers Polym.* **2008**, *11*, 593–607. [CrossRef]
76. Cao, L.; Liu, X.; Na, H.; Wu, Y.; Zheng, W.; Zhu, J. How a bio-based epoxy monomer enhanced the properties of diglycidyl ether of bisphenol A (DGEBA)/graphene composites. *J. Mater. Chem. A* **2013**, *1*, 5081–5088. [CrossRef]
77. Shibata, M.; Nakai, K. Preparation and properties of biocomposites composed of bio-based epoxy resin, tannic acid, and microfibrillated cellulose. *J. Polym. Sci. Part B Polym. Phys.* **2010**, *48*, 425–433. [CrossRef]
78. Kumar, P.P.; Paramashivappa, R.; Vithayathil, P.J.; Rao, P.V.S.; Rao, A.S. Process for isolation of cardanol from technical cashew (*Anacardium occidentale* L.) Nut shell liquid. *J. Agric. Food Chem.* **2002**, *50*, 4705–4708. [CrossRef]
79. Voirin, C.; Caillol, S.; Sadavarte, N.V.; Tawade, B.V.; Boutevin, B.; Wadgaonkar, P.P. Functionalization of cardanol: towards biobased polymers and additives. *Polym. Chem.* **2014**, *5*, 3142–3162. [CrossRef]
80. Maffezzoli, A.; Calò, E.; Zurlo, S.; Mele, G.; Tarzia, A.; Stifani, C. Cardanol based matrix biocomposites reinforced with natural fibres. *Compos. Sci. Technol.* **2004**, *64*, 839–845. [CrossRef]
81. Jaillet, F.; Darroman, E.; Ratsimihety, A.; Auvergne, R.; Boutevin, B.; Caillol, S. New biobased epoxy materials from cardanol. *Eur. J. Lipid Sci. Technol.* **2014**, *116*, 63–73. [CrossRef]
82. Darroman, E.; Durand, N.; Boutevin, B.; Caillol, S. New cardanol/sucrose epoxy blends for biobased coatings. *Prog. Org. Coat.* **2015**, *83*, 47–54. [CrossRef]
83. Atta, A.M.; Al-Hodan, H.A.; Hameed, R.S.A.; Ezzat, A.O. Preparation of green cardanol-based epoxy and hardener as primer coatings for petroleum and gas steel in marine environment. *Prog. Org. Coat.* **2017**, *111*, 283–293. [CrossRef]
84. Mooibroek, H.; Cornish, K. Alternative sources of natural rubber. *Appl. Microbiol. Biotechnol.* **2000**, *53*, 355–365. [CrossRef] [PubMed]
85. Baker, C.S.L.; Gelling, I.R.; Newell, R. Epoxidized Natural Rubber. *Rubber Chem. Technol.* **1985**, *58*, 67–85. [CrossRef]
86. Balakrishnan, H.; Nematzadeh, N.; Wahit, M.U.; Hassan, A.; Imran, M. Epoxidized natural rubber toughened polyamide 6/organically modified montmorillonite nanocomposites. *J. Thermoplast. Compos. Mater.* **2014**, *27*, 395–412. [CrossRef]
87. Hashim, A.S.; Ong, S.K. Study on polypropylene/natural rubber blend with polystyrene-modified natural rubber as compatibilizer. *Polym. Int.* **2002**, *51*, 611–616. [CrossRef]
88. Yoksan, R. Epoxidized Natural Rubber for Adhesive Applications. *Kasetsart J. (Nat. Sci.)* **2008**, *42*, 325–332.
89. Grande, A.M.; Rahaman, A.; Landro, L.D.; Penco, M.; Spagnoli, G. Self Healing of Blends Based on Sodium Salt of Poly(Ethylene-*co*-Methacrylic Acid)/Poly(Ethylene-*co*-Vinyl Alcohol) and Epoxidized Natural Rubber Following High Energy Impact. In Proceedings of the 3rd International Conference on Self-Healing Materials, Bath, UK, 27–29 June 2011.
90. Arroyo, M.; López-Manchado, M.A.; Valentín, J.L.; Carretero, J. Morphology/behaviour relationship of nanocomposites based on natural rubber/epoxidized natural rubber blends. *Compos. Sci. Technol.* **2007**, *67*, 1330–1339. [CrossRef]
91. Greve, H.-H. Rubber, 2. Natural. *ULLMANN'S Encycl. Ind. Chem.* **2012**, *31*, 583–594.
92. Hamzah, R.; Bakar, M.A.; Khairuddean, M.; Mohammed, I.A.; Adnan, R. A structural study of epoxidized natural rubber (ENR-50) and its cyclic dithiocarbonate derivative using NMR spectroscopy techniques. *Molecules* **2012**, *17*, 10974–10993. [CrossRef] [PubMed]
93. Mathew, V.S.; George, S.C.; Parameswaranpillai, J.; Thomas, S. Epoxidized natural rubber/epoxy blends: Phase morphology and thermomechanical properties. *J. Appl. Polym. Sci.* **2014**, *131*, 1–9. [CrossRef]

94. Imbernon, L.; Oikonomou, E.K.; Norvez, S.; Leibler, L. Chemically crosslinked yet reprocessable epoxidized natural rubber via thermo-activated disulfide rearrangements. *Polym. Chem.* **2015**, *6*, 4271–4278. [CrossRef]

95. Pire, M.; Norvez, S.; Iliopoulos, I.; Rossignol, B. Le Leibler, L. Epoxidized natural rubber/dicarboxylic acid self-vulcanized blends. *Polymer* **2010**, *51*, 5903–5909. [CrossRef]

96. Pire, M.; Norvez, S.; Iliopoulos, I.; Le Rossignol, B.; Leibler, L. Dicarboxylic acids may compete with standard vulcanisation processes for crosslinking epoxidised natural rubber. *Compos. Interfaces* **2014**, *21*, 45–50. [CrossRef]

97. Imbernon, L.; Pire, M.; Oikonomou, E.K.; Norvez, S. Macromol. Chem. Phys. 7/2013. *Macromol. Chem. Phys.* **2013**, *214*, 745. [CrossRef]

98. McCarthy, J.L.; Islam, A. Lignin Chemistry, Technology, and Utilization: A Brief History. In *Lignin: Historical, Biological, and Materials Perspectives*; American Chemical Society: Washington, DC, USA, 2000.

99. Chung, H.; Washburn, N.R. *Extraction and Types of Lignin*; Elsevier Inc.: Amsterdam, The Netherlands, 2015; ISBN 9780323355667.

100. Lora, J.H.; Glasser, W.G. Recent industrial applications of lignin: A sustainable alternative to nonrenewable materials. *J. Polym. Environ.* **2002**, *10*, 39–48. [CrossRef]

101. Ferdosian, F.; Yuan, Z.; Anderson, M.; Xu, C. Synthesis and characterization of hydrolysis lignin-based epoxy resins. *Ind. Crops Prod.* **2016**, *91*, 295–301. [CrossRef]

102. Ferdosian, F.; Yuan, Z.; Anderson, M.; Xu, C.C. Chemically modified lignin through epoxidation and its thermal properties. *J-FOR* **2012**, *2*, 11–15.

103. Asada, C.; Basnet, S.; Otsuka, M.; Sasaki, C.; Nakamura, Y. Epoxy resin synthesis using low molecular weight lignin separated from various lignocellulosic materials. *Int. J. Biol. Macromol.* **2015**, *74*, 413–419. [CrossRef] [PubMed]

104. Fache, M.; Auvergne, R.; Boutevin, B.; Caillol, S. New vanillin-derived diepoxy monomers for the synthesis of biobased thermosets. *Eur. Polym. J.* **2015**, *67*, 527–538. [CrossRef]

105. Fache, M.; Darroman, E.; Besse, V.; Auvergne, R.; Caillol, S.; Boutevin, B. Vanillin, a promising biobased building-block for monomer synthesis. *Green Chem.* **2014**, *16*, 1987–1998. [CrossRef]

106. Wang, S.; Ma, S.; Xu, C.; Liu, Y.; Dai, J.; Wang, Z.; Liu, X.; Chen, J.; Shen, X.; Wei, J.; Zhu, J. Vanillin-Derived High-Performance Flame Retardant Epoxy Resins: Facile Synthesis and Properties. *Macromolecules* **2017**, *50*, 1892–1901. [CrossRef]

107. Shibata, M.; Ohkita, T. Fully biobased epoxy resin systems composed of a vanillin-derived epoxy resin and renewable phenolic hardeners. *Eur. Polym. J.* **2017**, *92*, 165–173. [CrossRef]

108. Zhang, J. *Rosin-Based Chemicals and Polymers*, 1st ed.; Smithers Rapra Technology Ltd.: Shrewsbury, UK, 2012; ISBN 978-1-84735-506-5.

109. Liu, X.Q.; Huang, W.; Jiang, Y.H.; Zhu, J.; Zhang, C.Z. Preparation of a bio-based epoxy with comparable properties to those of petroleum-based counterparts. *eXPRESS Polym. Lett.* **2012**, *6*, 293–298. [CrossRef]

110. Liu, X.; Xin, W.; Zhang, J. Rosin-based acid anhydrides as alternatives to petrochemical curing agents. *Green Chem.* **2009**, *11*, 1018–1025. [CrossRef]

111. Deng, L.; Ha, C.; Sun, C.; Zhou, B.; Yu, J.; Shen, M.; Mo, J. Properties of Bio-based Epoxy Resins from Rosin with Different Flexible Chains. *Ind. Eng. Chem. Res.* **2013**, *52*, 13233–13240. [CrossRef]

112. Li, C.; Liu, X.; Zhu, J.; Zhang, C.; Guo, J. Synthesis, Characterization of a Rosin-based Epoxy Monomer and its Comparison with a Petroleum-based Counterpart. *J. Macromol. Sci. Part A* **2013**, *50*, 321–329. [CrossRef]

113. HEXCEL HexFlow®RTM 6 Product Data Sheet. Available online: https://www.hexcel.com/user_area/content_media/raw/HexFlow_RTM6_DataSheet.pdf (accessed on 1 October 2018).

114. Raquez, J.-M.; Deléglise, M.; Lacrampe, M.-F.; Krawczak, P. Thermosetting (bio)materials derived from renewable resources: A critical review. *Prog. Polym. Sci.* **2010**, *35*, 487–509. [CrossRef]

115. Ganesan, A.; Muthusamy, S. Mechanical properties of high temperature cyanate ester/BMI blend composites. *Polym. Compos.* **2009**, *30*, 782–790. [CrossRef]

116. Wang, J.; Liang, G.; Zhu, B. Modification of Cyanate Resin by Nanometer Silica. *J. Reinf. Plast. Compos.* **2007**, *26*, 419–429. [CrossRef]

117. Yuan, L.; Gu, A.; Liang, G.; Zhang, Z. Microcapsule-modified bismaleimide (BMI) resins. *Compos. Sci. Technol.* **2008**, *68*, 2107–2113. [CrossRef]

118. Nikafshar, S.; Zabihi, O.; Hamidi, S.; Moradi, Y.; Barzegar, S.; Ahmadi, M.; Naebe, M. A renewable bio-based epoxy resin with improved mechanical performance that can compete with DGEBA. *RSC Adv.* **2017**, *7*, 8694–8701. [CrossRef]

119. Dicker, M.P.M.; Duckworth, P.F.; Baker, A.B.; Francois, G.; Hazzard, M.K.; Weaver, P.M. Green composites: A review of material attributes and complementary applications. *Compos. Part A Appl. Sci. Manuf.* **2014**, *56*, 280–289. [CrossRef]

MDPI

St. Alban-Anlage 66

4052 Basel

Switzerland

Tel. +41 61 683 77 34

Fax +41 61 302 89 18

www.mdpi.com

Aerospace Editorial Office

E-mail: aerospace@mdpi.com

www.mdpi.com/journal/aerospace

CPSIA information can be obtained
at www.ICGtesting.com
Printed in the USA
BVHW021021020519

547198BV00021B/1728/P